Progress in
Colloid & Polymer Science

Editors: H.-G. Kilian (Ulm) and A. Weiss (Munich)

Surfactants, Adsorption, Surface Spectroscopy and Disperse Systems

Guest Editors:
B. Lindman (Lund), G. Olofsson (Lund),
and P. Stenius (Stockholm)

Springer-Verlag Berlin Heidelberg GmbH

ISBN 978-3-662-16084-8   ISBN 978-3-7985-1699-1 (eBook)
DOI 10.1007/978-3-7985-1699-1
ISSN 0340-255 X

© 1985 by Springer-Verlag Berlin Heidelberg
Originally published by Dr. Dietrich Steinkopff Verlag GmbH & Co. KG, Darmstadt in 1985
Softcover reprint of the hardcover 1st edition 1985

Copy editing: Cynthia Feast; Production: Holger Frey

Type-Setting: K + V Fotosatz GmbH, Beerfelden

Foreword

The present volume of Progress in Colloid and Polymer Science contains papers presented at the 8th Scandinavian Symposium on Surface Chemistry which was held at the University of Lund, Sweden, on June 4–6, 1984.

The Symposium was attended by about 160 scientists, of whom more than 50 represented the industrial community. The 50 non-Nordic participants gave the Symposium an international character. The program consisted of six plenary lectures, 21 oral and 38 poster presentations.

The main topics for the Symposium were: "Surfactant Systems", "Colloid Dispersions" and "Adhesion to Solid Surfaces". Both theoretical and applied aspects were covered. The last half-day session, devoted to "Microemulsions", was held jointly with the International Workshop on Microemulsions which was arranged in connection with the Symposium.

A special "Aniansson Memorial Session" was arranged on the afternoon of June 4 to commemorate the late Professor Gunnar Aniansson, University of Göteborg, Sweden. A posthumous paper by Aniansson on micellar kinetics was presented at the Symposium and it opens the present volume.

Financial support given to the Symposium by the Swedish Board for Technical Development is gratefully acknowledged.

Björn Lindman
Gerd Olofsson
Per Stenius

Contents

Gunnar Aniansson in memoriam

Professor Gunnar Aniansson died unexpectedly on January 2, 1984. He was a highly original thinker, deeply interested not only in physics, chemistry and other sciences, but also in the humanities and religious and moral questions.

Gunnar Aniansson was born in Södermanland on January 31, 1924. He graduated from the Royal Institute of Technology, Stockholm, in 1948. His doctoral thesis, "Retardation of α-particles in liquids" was defended in 1961 and in that year he also became "docent" in physical chemistry. In 1963 he was appointed to the new chair in physical chemistry at the University of Göteborg.

Gunnar Aniansson is probably best known for his work on the kinetics and dynamics of micelles. His theoretical paper from 1974 (together with Staffan Wall) on the kinetics of micelle formation marks a turning point since it provided the necessary theoretical background for the correct interpretation of apparently conflicting experimental results from various relaxation methods. Gunnar Aniansson made several important contributions in this field — an earlier unpublished paper is included in this volume. For this work he was awarded the Norblad-Ekstrand medal by the Swedish Chemical Society in 1983.

Gunnar Aniansson had already earlier made important contributions to surface chemistry. In 1950 and 1951 he published two papers (G. A. and O. Lamm (1950), Nature 165:357 and G. A. (1951), J Phys Colloid Chem 55:1286) on the determination of surface excesses by a radioactive method using, for example, sodium hexadecylsulfate labelled with ^{35}S. This work gave a direct and precise verification of Gibbs adsorption law.

The use of radioactive techniques to determine surface excesses was further extended in a series of elegant works, where the limited range (~ 1000 Å) of α-recoiling atoms in water was utilized. Surface excesses of Bi-212 ions were determined — e.g. coadsorbed with dodecylsulfate — by collecting and counting the recoiling T1-208 atoms above the surface.

In his doctoral thesis Gunnar Aniansson presented very accurate measurements of the stopping ranges of $5 \cdot 3$ MeV α-particles in liquids, using a new method and apparatus. He obtained such a high precision that he seriously considered developing a spectroscopic method based on the determination of stopping range distributions.

As a young professor at the new Institute in Göteborg, Gunnar Aniansson initiated research in several very demanding areas of physical chemistry and chemical physics. For example, a molecular beams group was formed, which is now led by docent Leif Holmlid and is strong and active both experimentally and theoretically.

An emphasis on a combination of theoretical and experimental work, with depth and skill in both, was characteristic of Gunnar Aniansson and made him a rather demanding research supervisor. He was an eminently good lecturer, in particular on the theoretical basis of physical chemistry, and eager to provide his students with a sound theoretical background.

Gunnar was fond of nature and outdoor activities and also enjoyed playing badminton with colleagues and students. His strongest interest in later years, however, was in religion and ethical questions. He made several journeys to Israel and stayed for extended periods at a monastery in Latroun, where he found a harmonic, peaceful and yet disciplined life well suited to deep thinking, fruitful both scientifically and for his inner peace.

For his pupils and coworkers Gunnar Aniansson was a friend, not always easy to cope with but usually charming, humorous and warm, and always trying to probe the depths of a problem and find the ultimate truth.

Mats Almgren

The mean lifetime of a micelle

G. E. A. Aniansson*

Department of Physical Chemistry, University of Göteborg and Chalmers University of Technology, Göteborg (Sweden)

Abstract: Different average lifetimes in micellar systems have been calculated on the basis of a simple stepwise mechanism, $A_{s-1} + A_1 \rightleftarrows A_s$. Expressions of the average residence time at a given aggregation number and the lifetime of a micelle are given.

Key words: Micelle, kinetics

Introduction

In kinetic measurements of micellization processes one generally obtains two relaxation times from which rate constants, the width of the micelle size distribution and information on the rarest intermediate single micelles can be deduced [1, 2]. From such measurements it is also possible, as we shall see, to calculate the mean lifetime of a micelle. Such a quantity is of course of interest in itself but further interest arises in connection with the role of the micelle as host for one or more solubilized or adsorbed molecules.

The kinetics of micellization has so far been successfully interpreted on the assumption that the fundamental process is a stepwise one, namely,

$$A_{s-1} + A_1 \underset{k_s^-}{\overset{k_s^+}{\rightleftarrows}} A_s \tag{1}$$

where A_s denotes an aggregate with aggregation number s. A particular micelle will then execute a random walk along the aggregation axis. For one with aggregation number s the probability for a step to the left in unit time will be k_s^- and for a step to the right $k_{s+1}^+ A_1$ where A_1 also denotes the concentration of the free monomer. Outside of equilibrium there will occur a net diffusive flow J_s from aggregation number $s-1$ to s which amounts to

$$J_s = k_s^+ A_{s-1} A_1 - k_s^- A_s. \tag{2}$$

Introducing the relative deviations of the concentrations A_s from the equilibrium values $\bar{A}_s$

$$\xi_s = \frac{A_s - \bar{A}_s}{\bar{A}_s} \tag{3}$$

and utilizing the fact that

$$k_s^+ \bar{A}_1 \bar{A}_{s-1} - k_s^- \bar{A}_s = 0 \tag{4}$$

the expression for the flow becomes

$$J_s = -k_s^- \bar{A}_s (\xi_s - \xi_{s-1} - \xi_1 - \xi_1 \xi_{s-1}) . \tag{5}$$

When ξ_1 is zero this expression is closely similar to Fick's first law for diffusion in a tube with varying cross section $A(x)$:

$$J(x) = -DA(x) \frac{\partial c}{\partial x}. \tag{6}$$

It is apparent that k_s^- corresponds to D, $\bar{A}_s$ to $A(x)$ and $\xi_s - \xi_{s-1}$ to $\partial c / \partial x$.

The concept of "lifetime" has not yet been defined. It could in principle have several possible meanings, two of which will be discussed here. The second to be treated is possibly the one that most naturally presents itself.

The residence time

The simplest concept of lifetime is the time that the micelle stays completely intact, that is the residence

---

* Died January 2, 1984

time at a given aggregation number. At macroscopic equilibrium – the only case for which we shall evaluate expressions – the probability in unit time that the micelle at s loses or gains a monomer and thus leaves its aggregation size is $k_s^- + k_{s+1}^+ \bar{A}_1$. The probability that the micelle stays at s more than t seconds will be $\exp[-(k_s^- + k_{s+1}^+ \bar{A}_1)t]$ and the mean residence time $(k_s^- + k_{s+1}^+ \bar{A}_1)^{-1}$.

At the maximum of the micelle distribution $s = \hat{n}$ one finds from (4) that $k_{\hat{n}}^+ \bar{A}_1 = k_{\hat{n}}^-$. It may reasonably be assumed [1, 2] that k_s^- and k_s^+ do not vary drastically with s in the micellar region so that we may put $k_{\hat{n}+1}^+ \cong k_{\hat{n}}^+$. The mean residence time at the most probable micelle size will then be very nearly $(2k_{\hat{n}}^-)^{-1}$. For a not very unsymmetrical micelle distribution $\hat{n} \cong n$, the mean aggregation number, so that as a representative value of the mean residence time in the micellar region we may take $(2k_n^-)^{-1}$. For sodium dodecyl sulphate at 25 °C the value for k_n^- obtained in measurements [2] is $1.0 \cdot 10^7$ seconds^{-1} which would yield a mean residence time of $5 \cdot 10^{-8}$ seconds.

Transition times

The mean time taken for a micelle to move from one aggregation number s'' to another, s' might be termed a transition time. It can be calculated in the following way [3]. Assume that micelles are added at s'' and removed at s' at the rate of J per unit time while A_1 is kept at $\bar{A}_1$. When the process is stationary the total number N of micelles on their way from s'' to s' will be

$$N = \int_0^\infty J\,dt \int_t^\infty f(t')\,dt', \tag{7}$$

where $f(t)\,dt$ is the fraction that arrives between t and $t + dt$.

Changing the order of integration one finds

$$N = J \int_0^\infty dt' \int_0^{t'} f(t')\,dt = J \int_0^\infty dt'\, t' f(t') \equiv J\langle t \rangle \tag{8}$$

where $\langle t \rangle$ is the mean transition time. Thus

$$\langle t \rangle = \frac{N}{J}. \tag{9}$$

In order to calculate N at given J we note that for $s' < s \leqslant s''$

$$-J = J_s = -k_s^- \bar{A}_s(\xi_s - \xi_{s-1}). \tag{10}$$

Since the monomer concentration is the equilibrium one, i.e. $\xi_1 = 0$.

Since all particles arriving at s' are removed we shall have $A_{s'} = 0$, or $\xi_{s'} = -1$. We further choose J such that $A_{s''} = \bar{A}_{s''}$, or $\xi_{s''} = 0$.

Dividing equation (10) by $k_s^- \bar{A}_s$ and summing both sides from $s' + 1$ to s'' we find

$$\xi_{s''} - \xi_{s'} = J \sum_{s'+1}^{s''} \frac{1}{k_s^- \bar{A}_s} \tag{11}$$

or

$$J = 1/R(s', s'') \tag{12}$$

where we have introduced

$$R(s', s'') \equiv \sum_{s'+1}^{s''} \frac{1}{k_s^- \bar{A}_s}. \tag{13}$$

If we perform the summation instead from $s+1$, $s' < s < s''$, to s'' we obtain

$$-\xi_s = J R(s, s'') \tag{14}$$

with an obvious extension of the notation (13). Thus

$$\xi_s = -J R(s, s'') = -\frac{R(s, s'')}{R(s', s'')}. \tag{15}$$

From the fact that all micelles arriving at $s' < s''$ are removed it follows that $A_s = 0$ for $2 \leqslant s \leqslant s'$ and hence that there is no contribution to N from this region. For $s > s''$ we note that $J_s = 0$ which means that in this region $\xi_s = \xi_{s''} = 0$ or that $A_s = \bar{A}_s$.

We then have

$$\begin{aligned} N &= \sum_{s>s'} A_s = \sum_{s>s'} (1 + \xi_s)\bar{A}_s \\ &= \sum_{s'+1}^{s''} \left[1 - \frac{R(s, s'')}{R(s', s'')}\right] \bar{A}_s + \sum_{s>s''} \bar{A}_s \\ &= \sum_{s'+1}^{s''} \frac{R(s', s)}{R(s', s'')} \bar{A}_s + \sum_{s>s''} \bar{A}_s. \end{aligned} \tag{16}$$

Combining equations (9), (12) and (16) we finally obtain

$$\langle t \rangle = \sum_{s'+1}^{s''} R(s', s)\bar{A}_s + R(s', s'') \sum_{s>s''} \bar{A}_s. \tag{17}$$

The lifetime of a micelle

The micelle size distribution in a typical surfactant solution above c.m.c. is something like that of the figure. In measurable amounts there are monomers and aggregates (micelles) in a distribution around an average aggregation number $n = \sum s \bar{A}_s / \sum \bar{A}_s$. Somewhere between the monomers and the proper micelles the distribution will normally have a very deep minimum. This portion is sketched greatly magnified in the figure. Since k_s^- is not expected to vary drastically in the intermediate region the contributions to $R(s', s'')$ will be mainly determined by the magnitude of $\bar{A}_s$, $s' + 1 \leqslant s \leqslant s''$. These contributions will be very large in the minimum and in a certain region around it but otherwise very small. This region is indicated by the limits s_1' and s_2' in the figure. The lower boundary of the region where $\bar{A}_s$ is non-negligible is denoted by s_2''. It would normally be larger than s_2'.

In calculating the lifetime of a micelle we will be interested in s'' values larger than s_2'' and s' values smaller than s_1'. Regarding the first sum in equation (17) we see that $R(s', s)$ will be negligible for $s < s_1'$. Between s_1' and s_2' $R(s', s)$ will increase towards a value $R = R(s_1', s_2')$ but $\bar{A}_s$ will be negligible and will remain so until $s = s_2''$. $R(s', s)$ will increase with negligible amounts at $s > s_2''$ and will therefore stay practically constant equal to R. Equation (17) can therefore be written

$$\langle t \rangle = \sum_{s_2'}^{s''} R \bar{A}_s + R \sum_{s > s''} \bar{A}_s = R \sum_{s \geqslant s_2''} \bar{A}_s = R C_3 \quad (18)$$

where C_3 is the total number of micelles. Within the approximations used the lifetime of the proper mi-

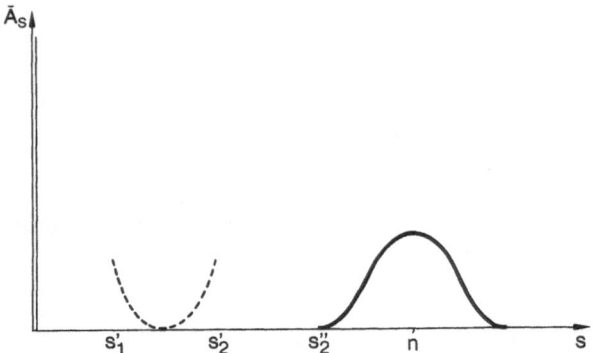

Fig. 1. The number of species with aggregation number s as a function of s, full line. Broken line: Strongly magnified portion bounded by s_1' and s_2' around the distribution minimum

celles ($s \geqslant s_2''$) is independent of the micelle size. From the argument it is also clear that $\langle t \rangle$ is also essentially independent of s' as long as it is smaller than s_1'.

This behaviour can be understood in the following way. A particular micelle will undergo a random walk along the aggregation axis with a probability per unit time for a step to the left equal to k_s^- and for a step to the right equal to $k_{s+1}^+ \bar{A}_1$. The ratio of these two is

$$\frac{k_s^-}{k_{s+1}^+ \bar{A}_1} \cong \frac{k_s^-}{k_s^+ \bar{A}_1} = \frac{\bar{A}_{s-1}}{\bar{A}_s} . \quad (19)$$

Above the maximum in the distribution this will be increasingly larger than unity and between the maximum and the minimum it will be smaller than unity. Below the minimum it will again be larger than unity. When the micelle is in the proper micellar region there will then be a larger probability for a step towards the distribution maximum than away from it. (This is what holds the distribution together.) In order to arrive at the minimum it must in succession make a large number of steps to the left against the probability and this will occur very seldom compared to a random walk to and from around the distribution maximum. On average the micelle will have made many such tours before arriving past the minimum and becoming dissolved and it will thus make little difference exactly from where in the proper micellar region it started out.

When it has arrived below s_1', say, the probability for a step to the left will always be larger than one to the right and the aggregate will comparatively quickly arrive at $s = 1$ and be dissolved.

The quantity R in equation (18) is also contained in the expression for the relaxation time, τ_2, of the slow process holding well above the c.m.c.

$$\frac{1}{\tau_2} = \frac{n^2}{R \bar{A}_1 \left(1 + \dfrac{\sigma^2}{n} a \right)} . \quad (20)$$

σ^2 is the variance of the proper micellar distribution, and

$$a = \frac{A_{\text{tot}} - \bar{A}_1}{\bar{A}_1} = \frac{n C_3}{\bar{A}_1} \quad (21)$$

where A_{tot} is the total concentration of surfactant. Inserting R from equation (20) and using equation (21) one obtains

$$\langle t \rangle = \frac{n\tau_2 a}{1 + \dfrac{\sigma^2}{n}a}. \tag{22}$$

τ_2 sometimes changes by orders of magnitude with concentration [2], but as an example we may take sodium dodecyl sulphate at 20°C and about twice the c.m.c. Using experimental values [2] one finds for $\langle t \rangle$ the value 0.07 sec.

Except close to the c.m.c. $a/[1 + (\sigma^2/n)a]$ is the order of one so that generally the order of magnitude of $\langle t \rangle$ is determined by the product $n\tau_2$. Since n is typically of the order of one hundred the lifetime of a micelle is generally much longer than the relaxation time of the slow process.

References

1. Aniansson EAG, Wall SN (1974, 1975) J Phys Chem 78:1024 and 79:857
2. Aniansson EAG, Wall SN, Almgren M, Hoffmann H, Kielmann I, Ulbricht W, Zana R, Lang J, Tondre C (1976) J Phys Chem 80:905
3. v Smoluchowski M (1923) Abhandlungen über Brownsche Bewegung und verwandte Erscheinungen, Ostwald Klassiker Nr 207, Leipzig

Received June 6, 1984;
accepted November 20, 1984

Address for enquiries:

Dr. S. Wall
Department of Physical Chemistry
CTH/GU
S-41296 Göteborg, Sweden

The standard picture of ionic micelles

D. W. R. Gruen

Department of Applied Mathematics, Research School of Physical Sciences, Australian National University, Canberra, (Australia)

Abstract: A description of the state of amphiphilic molecules in ionic micelles is presented. This description is dubbed "the standard picture of ionic micelles" because the evidence supporting it is by now overwhelming.

A simple theoretical model for the packing of chains in amphiphilic aggregates is presented. The model is based on the standard picture. Aggregates are assumed to have symmetrical hydrophobic cores in which only the chains may exist. The model involves generating all possible internal bond sequences – *trans, gauche*$^+$ and *gauche*$^-$ at each bond – of a single amphiphilic molecule. The probabilities of these different conformations are constrained in such a way that, when an ensemble average is taken over all conformations, the hydrophobic core of the aggregate is packed at liquid hydrocarbon density throughout. To a limited extent the chains may also exist outside the core, in which case they incur a hydrophobic free energy cost. The model accurately reproduces the static properties of a recent molecular dynamics simulation of a bilayer containing 128 chains. For spherical micelles the model is in good agreement with neutron diffraction experiments which measure the mean position and freedom of movement of the terminal methyl groups. It is in excellent agreement with NMR T_1 relaxation experiments from which an order parameter down the amphiphile chain can be deduced.

Evidence which has been claimed to invalidate the standard picture of ionic micelles is examined in detail and found wanting.

Key words: Statistical chain model, dry micelle core order parameters

1. The standard picture of ionic micelles

To establish the standard picture of the state of amphiphilic molecules in ionic micelles I will draw on both theoretical and experimental evidence. I have no intention of providing an exhaustive description of all the evidence which supports the picture but I will point to key observations which are particularly telling. The standard picture is widely applicable. However, for simplicity and concreteness, we will consider an ionic amphiphile with a single fully saturated alkyl chain under solution conditions (e.g., low salt concentration) where the micelles which form are small globules. (The word globules is intentionally used so as not to prejudice the discussion with any particular shape.) We now marshall the evidence which will lead us to the standard picture.

Fully saturated alkyl chains are very flexible. Each C – C bond in a fully extended alkyl chain exists in a *trans* state but each bond can also exist in two other conformations – the *gauche*$^+$ and *gauche*$^-$ states which can each be formed at an energy of $0.8 – 1.0\,kT$ at room temperature [1]. Therefore, in a fully saturated alkyl chain of length (say) 8 or more, there are a large number of fairly low energy (and therefore accessible) states of the chain.

In forming a micelle, amphiphile chains aggregate in the micelle's core to reduce expensive hydrocarbon-water contact. It is possible to imagine that the chains could pack in a frozen array of parallel all-*trans* chains, and indeed this is the packing found in solid bulk *n*-alkane. However, without considerable expensive hydrocarbon-water contact, it is impossible to see how such an array could exist in a globular micelle. We should expect, therefore, that the interior of a micelle is liquid, and that the alkyl chains are conformationally disordered. This expectation is borne out by experiments, among the most definitive of which are measurements of NMR T_1 relaxation times of ^{13}C nuclei in the chains [2, 3]. These measurements reveal intrachain motions with correlation times around 10^{-11} s, a value in close agreement with the correlation times calculated from similar measurements on liquid *n*-alkanes.

Given the very low solubility of water in bulk *n*-alkane phases (about one water molecule for each $2 \times 10^4 CH_2$ groups [4]), we should expect negligible water penetration into the aggregate core. For bilayers, this expectation is strongly supported by neutron diffraction experiments on a lamellar phase [5], and by a comparison of capacitance and optical reflectance measurements on 'black' lipid films [6]. For micelles, claims of unequivocal experimental evidence for extensive water penetration in the core have been shown to be unfounded [7]. The experimental evidence suggesting minimal water penetration in micellar cores is now very strong indeed. It comes from measurements of free energies of transfer of nonpolar molecules and amphiphiles into micelles [8], viscosity and diffusion measurements [7], NMR relaxation measurements [9, 10] and neutron diffraction studies [11, 12]. We shall have reason to further examine the issue of water penetration in the core of micelles in section 3 of this paper.

Accepting the above arguments, the micellar core is a non-polar liquid consisting almost exclusively of fully saturated alkyl chains. The packing density of the chains is determined by the combined action of short-range interatomic repulsive forces (arising from the overlap of electron clouds) and longer range van der Waals attractions. As the chains are chemically identical to *n*-alkane chains, we should expect a packing density almost identical to that found for bulk *n*-alkane, and indeed, experiments confirm this expectation [13]*).

We may now summarize the standard picture of ionic micelles in three simple points:

(i) on average, each amphiphile associated with a micelle has almost all of its hydrocarbon chain in the micelle core;

(ii) the ionic headgroups, ions and water are almost totally excluded from the micelle core;

---

*) If the surface of the micelle is a surface of tension, there will be a Laplace pressure acting on the aggregate interior. However, for any reasonable value of the *net* surface tension acting at the micelle surface (say, $\gamma < 10$ mN m^{-1}), even for small spherical micelles (say, with a radius of 15 Å), this Laplace pressure leads to a negligible increase in density. Thus, assuming that the compressibility of the aggregate interior is the same as that of bulk dodecane [14] leads to an estimate for the increase in density of less than 0.6%.

(iii) the amphiphile chains (which cannot separate from their respective headgroups) are conformationally disordered (liquid-like) and fill the core at approximately liquid *n*-alkane density throughout.

In point (iii), the parenthetical clause is included because some interpretations [15, 16] of the 'liquid hydrocarbon droplet' micelle model seem to ignore it.

The standard picture implies that the chain-water interface is sharp (with an average roughness of the order of a few Å) and that the headgroups sit just outside the micelle core in a layer again with a roughness of the order of a few Å. In most respects, this picture is very similar to Hartley's classical description of ionic micelles [17]. The standard picture of ionic micelles will be defended in section 3. Part of the armoury for the defence is a quantitative model for the chains which will now be described.

2. A quantitative model for the chains in amphiphilic aggregates

The standard picture of ionic micelles forms a common starting point for three recent attempts at modelling the state of the chains in amphiphilic aggregates: Dill and Flory's lattice model [15, 16], Fromherz' surfactant-block model [18] and a model devised by the present author [19-21]. For each of these models, it is necessary to make further assumptions which are not part of the standard picture. These further assumptions are necessary in order to render the models soluble.

For the present model, it is necessary to assume that the amphiphilic aggregate core has a symmetrical shape (spherical, cylindrical or bilayer-shaped). All possible internal bond sequences (*trans, gauche*$^+$ and *gauche*$^-$ at each bond) of a *single* amphiphilic molecule are generated on a computer. The probabilities of these different conformations are constrained in such a way that, when an ensemble average is taken over all conformations, the core of the aggregate is packed at liquid hydrocarbon density throughout. Some chain conformations with part of the chain outside the hydrophobic core are allowed, but these conformations are subject to a hydrophobic free energy cost. Thus, a region of the aggregate is allowed (between the core and the pure aqueous phase) where the chains, headgroups and water coexist.

In this model, the single amphiphile chain is treated as accurately as possible. Well established values for the $C-C$ bond length, and for the relative energy and bond angles of *trans* and *gauche* conformers [1] are all included in the model. This approach was pioneered by Marcelja [22] for bilayers and extended to a spherical micelle by the present author [19]. Apart from a recent molecular dynamics simulation of a bilayer [23], *all other models* of the hydrophobic core of amphiphilic aggregates treat the chains much less realistically.

Comparison of the model with a molecular dynamics simulation of a bilayer

A stern test of our 'single-chain' model is to compare results derived from it with results derived from a recent molecular dynamics simulation [23]. This simulation *explicitly* considers the interactions of 128 G-$(CH_2)_8CH_3$ amphiphiles in a bilayer (G is a model headgroup). To date, this simulation is *by far* the most reliable attempt to model the short time and short distance properties of the chains in a bilayer. It is therefore of considerable interest to compare the results of this simulation with any prospective model of the interior of amphiphilic aggregates.

Figures 1–5 display comparisons between the simulation and the present model. Further comparisons are presented in [21]. It is clear that for a wide range of static properties, the two models produce remarkably similar results. Apart from setting the energy of *gauche* kinking equal in the two models, it should be emphasised that this agreement occurs without any freely adjustable parameters. Further, *none* of the many proposed models of the bilayer interior (apart from earlier single chain models [22, 25]) are capable of providing this sort of agreement. For example, if the Dill and Flory model was used to model this system, there must be $9/3.6 = 2.5$ (hence 2 or 3) 'segments' in the Dill-Flory chains because the segments must be equidimensional [15]. Such chains have only 1 or 2 internal bonds (in contrast to 7 in the real chains) and so the Dill and Flory model can only provide very rough information about the quantities displayed in figures 1–5.

Our single-chain model is a close relative of the rotational isomeric state or random-coil model used to describe properties of bulk liquid n-alkane [1]. The random-coil model assumes that the energy of a chain conformation is simply the intramolecular energy of *gauche* kinking. No account is taken of the local conformations of surrounding chains, nor of the fact that for some bond sequences the chain will pass through itself or its neighbours (i.e., excluded-volume effects are ignored). This extremely simple model cannot possibly account for all the static properties of n-alkanes. Indeed, wide angle X-ray scattering and depolarized Rayleigh scattering experiments indicate

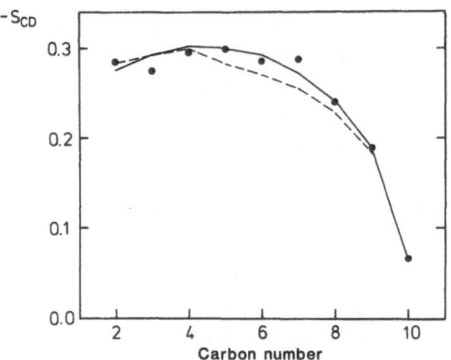

Fig. 1. C–D bond order parameters for different segments in the amphiphile chain. Carbon number 2 is the CH_2 (or CD_2) group bonded to the headgroup while number 10 is the terminal CH_3 group. $S_{CD} = \langle \frac{3}{2}\cos^2\theta - \frac{1}{2}\rangle$, where θ is the angle between the C–D bond vector and the bilayer normal. ●, experimental order parameters [24] down the decanoate chain in a 32 wt% sodium decanoate, 38 wt% decanol, 30 wt% water lamellar phase. – – –, from the molecular dynamics simulation. ———, from our single chain model. For the single chain model, the standard deviation of each point is 0.003. For reasons described elsewhere [21], the average area per amphiphile is slightly different in the two models (25 Å^2 for the molecular dynamics simulation; 25.6 Å^2 for the single chain model)

that there is some (but not much [26]) short-range order between liquid n-alkane chains (which the random-coil model implicitly ignores). Nevertheless, the random-coil model has been remarkably successful in predicting many properties of n-alkanes, even when these properties depend on the spatial distribution of

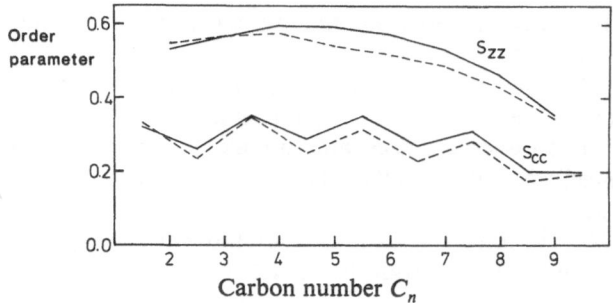

Fig. 2. Order parameters S_{zz} and S_{cc} along the chain. For carbon C_n, S_{zz} is the order parameter of the vector joining C_{n-1} to C_{n+1} with respect to the bilayer normal. It is the order parameter of the local axis of the chain. S_{cc} is the order parameter of the C–C bond with respect to the bilayer normal. ———, from the single chain model. – – –, from the molecular dynamics simulation. For the single chain model, the standard deviations for the S_{zz} data are 0.004 and for the S_{cc} data, 0.007

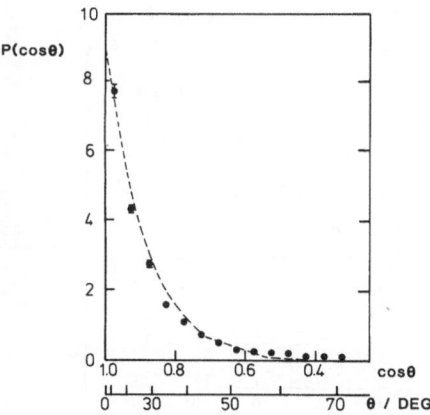

Fig. 3. Probability distribution of molecular tilt, $P(\cos\theta)$. Molecular tilt is defined by the vector joining the middle of the first to the middle of the sixth bond in the amphiphile. The tilt angle, θ, is then the angle between this vector and the bilayer normal. The data points and the dashed line display $P(\cos\theta)$ for the single chain model and the molecular dynamics simulation respectively

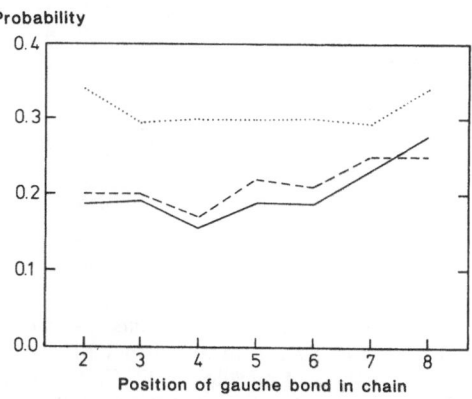

Fig. 5. The probability density of the position of *gauche* bonds in the chain. Bond 2 is the H_2C-CH_2 bond nearest the headgroup. Full line, dashed line and dotted line mark the results from the single chain model, the molecular dynamics simulation and the random coil model respectively. The results from the molecular dynamics simulations were quoted to two significant figures [23]. The standard deviation in the single chain results is 0.005

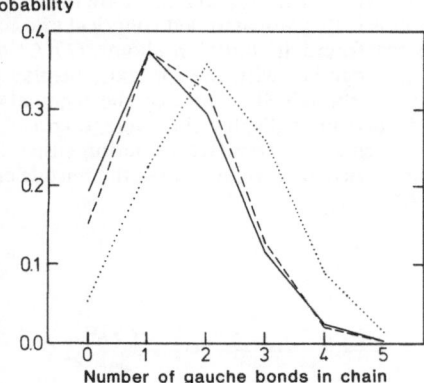

Fig. 4. The probability density of the number of *gauche* bonds in the chain. Full line, dashed line and dotted line mark the results from the single chain model, the molecular dynamics simulation and the random coil model (used to model bulk liquid hydrocarbon — see text) respectively. The standard deviation in the single chain results is 0.003

atoms less than 10 Å apart [1, 26–28]. Recent molecular dynamics simulations on bulk liquid *n*-butane [29] and *n*-hexane [30] also suggest that it accurately estimates the average proportion of *gauche* bonds.

It is interesting to speculate on why the random-coil model is so successful. Flory [1] presents strong reasons for believing in it. Further, an examination of the *intermolecular* carbon-carbon pair correlation

function in bulk *n*-octane (generated by molecular dynamics [31]) reveals almost no structure. Thus, the assumption that a single chain sits in a structureless soup which makes its van der Waals energy independent of its conformation (which is implicit in the random-coil model) may well be a good one.

Our single-chain model may be roughly described in these terms: one end of an (admittedly short) 'random coil chain' is anchored near the surface of the amphiphilic aggregate and the probabilities of the chain conformations are changed as little as possible consistent with setting up a state of constant density in the aggregate interior.

Since the random-coil model is a good model of bulk *n*-alkane, it is not so surprising that the present model is successful in reproducing a wide range of the static properties of the lipid bilayer molecular dynamics simulation.

Applying the model to spherical micelles

Having demonstrated the ability of the single chain model to mirror the behaviour of a large molecular dynamics simulation, it is time to confront the model with some quantitative experimental results on spherical micelles.

Figure 6 displays a comparison between the model and experimental results for the $C-H$ (or equivalently, $C-D$) bond order parameters as a function of position down the chains of roughly spherical dodecyl-

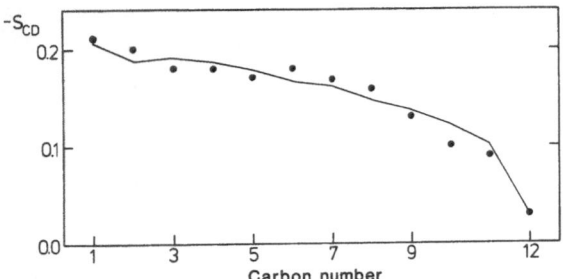

Fig. 6. Comparison of experimental and model $C - D$ order parameters (with respect to the local normal of the aggregate) for roughly spherical micelles at 40 °C. ●, results derived from T_1 NMR relaxation measurements on dodecyltrimethylammonium chloride (DOTAC) at 30 wt% by Walderhaug et al. [3]. ———, model results assuming a spherical hydrophobic core. The standard deviation of the model results is 0.007

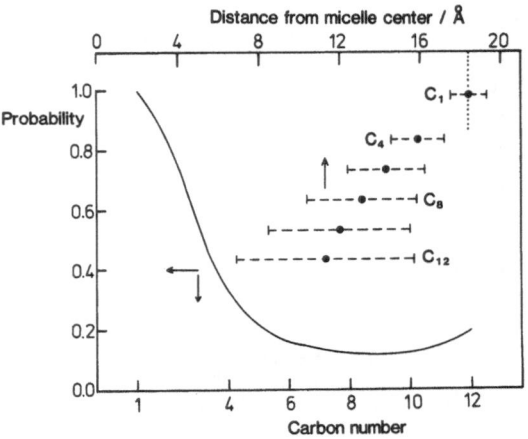

Fig. 7. Probability of each segment in the chain being at least partly in contact with water and the distribution of position of chain segments in a spherical micelle containing dodecyl chains with an aggregation number 78. The full line (together with the left-hand and bottom axes) displays the probability of different segments being at least partly in contact with water. To be deemed at least partly in contact with water, the C atom in CH_2 segments must be outside the micelle core or within 1.86 Å of the core surface. The terminal CH_3 group is deemed in contact with water if it is within 2.35 Å of the core surface. 1.86 Å and 2.35 Å are the radii of the respective groups if they are assumed spherical and have the partial volumes found in liquid *n*-alkane (27 Å^3 and 54 Å^3). The points, together with the top axis, display the mean distances from the micelle centre of the segments in the chain. The dashed lines display the average spread of position of the segments (one standard deviation either side of the mean). The vertical dotted line marks the micelle core surface (at 18.4 Å)

trimethylammonium chloride micelles. The model fits the experimental results very closely but it should be noted that two parameters are used to generate the model results. (The parameters are: the micelle core radius which takes the value 19.2 Å and a parameter which determines the relative weighting of different headgroup orientations.)*)

The micelle chains show a tendency to order perpendicular to the micelle surface, though this tendency becomes progressively weaker as the free end of the chain is approached (see figure 6). In this respect, the chains behave quite similarly to the chains in a bilayer (see figure 1). In fact, the model demonstrates that at the same mean area per chain, the order parameter profile along the chain is very similar whether the aggregate is a bilayer, a cylinder or a sphere [21].

Figures 7 – 9 display results for a spherical micelle containing 78 G-$(CH_2)_{11}CH_3$ amphiphiles (G is a model headgroup) at 35 °C. These conditions are chosen to mirror a recent neutron diffraction study [11]. The model spherical micelle core has a radius of 18.4 Å, which implies the existence of a hole at the centre of the micelle with a volume of 21 Å or 40% of the volume of a CH_3 group. Of course, real micelles exist in an equilibrium distribution of shapes, and all nonspherical shapes have some parts of the micelle surface nearer to the micelle centre than for a sphere of the same volume. Even very small deviations away

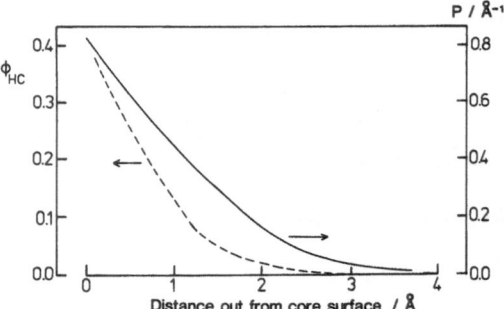

Fig. 8. Volume fraction filled by hydrocarbon chains, ϕ_{HC} (– – –) (left-hand axis) and probability distribution of headgroups, P (———) (right-hand axis) as a function of distance out from the micelle core surface. The results are derived for the micelle containing dodecyl chains with a core radius of 18.4 Å. The mean position of the headgroup is 1.0 Å out from the core surface while its standard deviation of position is 0.8 Å. All these results are almost identical for a micelle with dodecyl chains and a core radius equal to a fully extended chain (16.7 Å)

---

*) The fully extended length of a dodecyl chain is ~16.7 Å and hence a micelle with a perfectly spherical core with radius 19.2 Å has a hole at its centre slightly larger than a single CH_3 group. This situation will be discussed shortly.

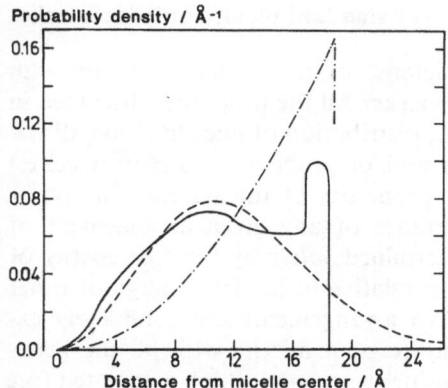

Fig. 9. Comparison of results derived from neutron diffraction experiments and from the model for the distribution of terminal methyl groups in spherical dodecylsulphate micelles at 35°C. The micelles are observed to have a mean aggregation number of 78. If the methyl group probability density is $p(r)$ where r is the distance from the micelle center, the plot shows the volume weighted probability density $P(r) = 4\pi C r^2 p(r)$, where C is a normalization constant. ———, model results; – – –, from the neutron diffraction experiments [11]; –·–·–, profile generated assuming that the CH_3 groups are distributed uniformly in a hydrophobic core containing all 78 chains and hence with radius 18.8 Å

from a perfect sphere would eliminate a hole of this volume.

Interestingly, our best estimate for the free energy cost of forming a hole with a volume of a CH_3 group is $1.9\,kT$ at room temperature [20] so the free energy cost of forming a hole with a volume of 21 Å^3 should be substantially less than $1\,kT$ (as van der Waals energy is strongly distance dependent). Because this energy is so small, we should expect the spherical shape to form a non-negligible part of the ensemble of equilibrium shapes of micelles even when the aggregation number is 78.

The model predicts that transfering amphiphiles from the random coil state (characteristic of a bulk liquid *n*-alkane environment) to this micelle entails a free energy cost of $0.5\,kT$ per amphiphile. Forming the hole at the micelle centre accounts for only ∼1% of this cost. The rest of the cost arises from the chains being forced to change the probabilities of their conformations in order to pack the micelle core at constant density. In the random coil state each amphiphile chain has an average of 3.67 *gauche* bonds, while in this micelle, each chain averages 3.06 *gauche* bonds. Of course, one or two chains must be completely straight in order to reach near the micelle centre, but *on average,* each chain loses about 0.6

gauche bonds on transfer to this micelle. It should be clear that in terms of the chain statistics, packing into this micelle involves only a slight perturbation from the random coil state.

Figure 7 displays some consequences of the fact that there are a large number of conformations accessible to the chains in this micelle while figure 8 shows the consequences of the strong hydrophobic effect. The core of the micelle has a volume equal to 74 $-(CH_2)_{11}CH_3$ chains while the aggregation number of the micelle is 78. On average therefore, 94% of the chain volume sits in the dry micelle core.

Figure 9 shows the complete spatial distribution of the terminal CH_3 groups in these spherical micelles. The model results are compared with results from a neutron diffraction study [11]. The model profile shows a pronounced peak near the micelle surface (though it should be noted that, if the CH_3 groups were distributed completely uniformly throughout the micelle, there would be ∼60% more of them at the surface than the model predicts − see figure 9). This peak occurs because volume increases sharply with distance from the centre of the micelle, and also because there are many conformations which go some way into the core and then return towards the surface. 14% of all the CH_3 groups sit within 1.5 Å of the model core surface, and of these almost half (44%) belong to chains which somewhere along their length have ventured at least 3 Å into the core.

The peak is particularly sharp because, in the model, CH_3 groups are completely confined to the micelle's hydrophobic core [20, 21]. Relaxing this restriction would broaden the peak and lower its height. For real micelles, polydispersity of size and fluctuations of shape must further broaden and lower the peak.

Notwithstanding these qualifications it is clear that the neutron diffraction results show no hint of the peak. However, these results were obtained only for momentum transfers with a magnitude $Q \lesssim 0.24$ Å^{-1} (and hence $QR < 4.6$ where R is the average radius of the micelles). Such measurements are only sensitive to features of the methyl group distribution with low spatial frequencies. This point is graphically illustrated by Cabane et al. [12] who generate the scattering curves for model spherical particles of the same average size but formed of concentric spherical shells with a variety of scattering densities. For the four different scattering density profiles considered by Cabane et al., the resulting scattering curves are all identical for $QR < 4.6$ (see Figs. 2 and 4 in [12]). Bendedouch et al.'s neutron diffraction results over this range of QR values are well fitted by assuming a

Gaussian distribution for the CH_3 groups [11]. At this level of approximation, the model gives close agreement with the experiments for both the mean $\langle r \rangle$ and standard deviation (σ) of the distance from the micelle centre to the CH_3 groups. Thus, the model gives $\langle r \rangle = 11.3$ Å and $\sigma = 4.4$ Å while the experiments yield $\langle r \rangle = 11.9$ Å and $\sigma = 5.0$ Å. Unfortunately, this experiment is not a very stern test of the theory because at this level of approximation a very simple model based directly on the standard picture developed in section 1 also gives good agreement with this experiment [21].

Figure 10 shows a scale drawing of a spherical dodecylsulphate micelle.

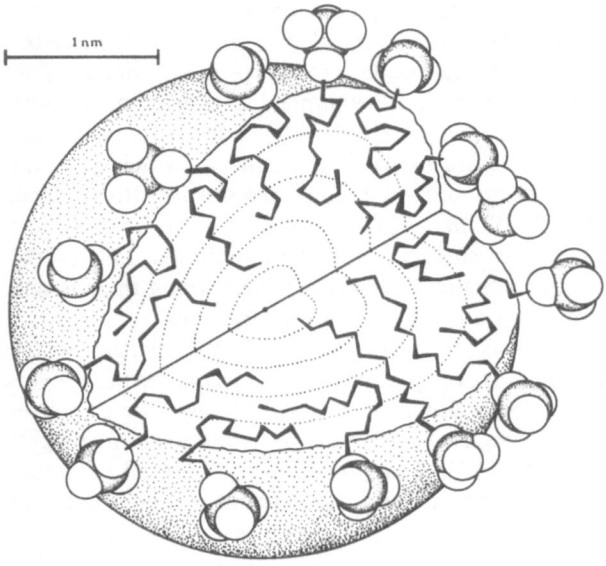

Fig. 10. A scale drawing of a spherical dodecylsulphate micelle. The hydrophobic core surface is drawn as a perfect sphere although there is, of course, an equilibrium distribution of core shapes. The hydrophobic core radius is equal to the fully extended length of the chain, 16.7 Å. At 25 °C, 55 chains can fit *inside* the core while the aggregation number of the micelle is approximately 58. The hydrophobic core is divided into five regions, each of thickness 3.34 Å. The model is used to dictate the number of chain segments belonging to each region. For clarity, only the $C - C$ skeletons of the chains are drawn. It appears that the outer parts of the micelle are crowded with chains while the central regions are almost empty. However, the partial volume of a CH_3 group is about twice that of a CH_2 group and also the figure is a two dimensional picture of an evenly packed three dimensional structure. By way of illustration, the innermost region has a volume of 156 Å^3; room for 2.9 CH_3 groups. On average, one of the fourteen amphiphiles shown should have its CH_3 group in this region only 70% of the time!

3. In defense of the standard picture of ionic micelles

I begin my defense of the standard picture with some general remarks. All the properties discussed in this section (e.g., distribution of micelle shape, distribution of water and/or probe molecules in micelles) are *equilibrium* properties of the system. The probability of occurrence of any given arrangement of molecules is determined *solely* by the free energy of that arrangement relative to the free energy of other arrangements. No arrangements are completely excluded but many (e.g., a micelle with all the headgroups on the inside) have such a high associated free energy that they contribute only infinitesimally to the properties of the system. We are usually only interested in those arrangements which contribute substantially to the ensemble of possibilities. These comments are particularly important for micelles because most measured properties of micelles represent an average over a large number of arrangements of the molecules. For example, in the model spherical micelle studied in figure 7, the terminal CH_3 group is, on average, closer to the micelle centre than any other group. It is also near the micelle surface more often than, say, C_8 (see figure 7). These two statements are consistent because there are many arrangements of the amphiphile which contribute substantially to the ensemble of possibilities.

We turn now to a discussion of dynamic measurements. Consider the fact that a monomer is associated with a sodium dodecylsulphate micelle for $10^{-6} - 10^{-5}$ s [32]. This is roughly a factor of 10^5 slower than intrachain motions (see earlier). Nevertheless, one might conclude that, in order for monomer exchange to be occurring at this rate, any instantaneous snapshot of the micelle must reveal many monomers 'half in and half out' of the micelle. In other words, one might conclude that the dynamics of exchange must impose a substantial 'dynamic roughness' on the micelle. Not so. Free energy considerations determine the dynamic properties of a system just as they determine the equilibrium properties. If free energy considerations require a fairly sharp hydrocarbon-water interface (and they do [21]), dynamics cannot overturn this requirement. If one wishes to discuss equilibrium averages in dynamic terms (something which complicates an already complex subject) one can do so. Thus, in the above example, for the *vast* majority of the $10^{-6} - 10^{-5}$ s a given monomer is associated with a micelle, it sits with almost all of its chain inside the aggregate's hydrophobic core and hence with its headgroup in close proximity to the core surface. It does this because of the prohibitive free energy cost of doing otherwise.

The standard picture presented in section 1 does not require that micellar cores be perfectly spherical. The model presented in section 2 assumes that they are in order to render the model soluble. The model is useful here because it enables us to make quantitative estimates which are invaluable when assessing arguments. However, use of the model should not be regarded as a tacit admission that micellar cores are perfect spheres!

The standard picture of the structure of ionic micelles is not universally accepted. Space does not permit a discussion of all points of disagreement but we can at least attempt to answer some of the criticisms advanced by one of the prominent opponents of the standard picture. Menger has advanced [33] and recently defended [34] an alternative view of ionic micelles based on the work of several groups [35 – 39]. "Menger micelles have a central hydrophobic core; this is surrounded by a much larger region composed of hydrocarbon, water, headgroups, and counterions. Disorganization is rampant, and the surface is rough. Chemists speak of 'degree of water penetration', but this is not really a precise phrase in the context of a Menger micelle. Penetration is deep in the sense that the core ... is relatively small in volume" [34].

Here we shall discuss *all* the arguments advanced by Menger and Doll [34] in favour of the Menger micelle. We begin with an analysis of the work by other groups [35 – 39] which Menger and Doll acknowledge as contributing to the formulation of the Menger micelle. We preface this analysis with some general remarks.

As has been stressed, the standard picture of ionic micelles admits the possibility of an equilibrium distribution of micelle core shapes. In contrast to the Menger micelle, the standard picture *requires* that at the surface of almost all these equilibrium shapes, the average volume fraction of hydrocarbon must fall from a value near one to a value near zero over a distance of the order of the linear dimensions of a water molecule or a CH_2 group (i.e., $1 – 3$ Å). Thus the standard picture implies that the core (which is almost exclusively composed of groups with low polarizabilities) *is in intimate molecular contact* with a region which is highly polar (containing headgroups, counterions and the surrounding aqueous phase). This situation is sufficiently different from any bulk *molecular solution* that one must not expect the latter to be a good model of the former either in terms of dielectric response or in terms of solvation properties. (To give but one simple example: a benzene molecule

by sitting at the micelle surface displaces hydrocarbon and hence replaces a small region of *n*-alkane chain-water contact with benzene-water contact. Although clearly not a surface-active molecule, the benzene spends a disproportionate fraction of time at the micelle surface because benzene-water contact is slightly favoured over *n*-alkane water contact [40]. Such an effect is not reproduced in a molecular solution of benzene in any solvent.)

In the light of these remarks, we comment on the experiments of de Mayo and Sydnes [35], Whitten et al. [38] and Zachariasse et al. [39]. All of the probe molecules (or groups) used by these workers will prefer the surface of the micelle to its interior. It is therefore not surprising that these probes experience a partly polar environment (see also [41, 42]). Given this non-ideal behaviour of the probes, these experiments *do not* provide any evidence contrary to the standard picture.

In passing we note that Whitten et al. observe a substantial difference in the behaviour of their stilbene derivatives S_4 to S_{16} depending on whether they are solubilized in micelles or vesicles. Whitten et al. conclude that in micelles the stilbene chromophore always spends a substantial fraction of time in contact with the aqueous phase while in vesicles it does not. We must be careful using the single chain model to interpret these results because of the perturbing effect of the stilbene chromophore. Nevertheless, it is interesting that the single chain model predicts that in a micelle all parts of the chain spend an appreciable fraction of time near the core surface (figure 7) but in a bilayer there is almost no contact with the surface of those segments in the chain beyond C_4 [20]. This is a consequence of the very different surface area/volume ratios for these two geometries. It is not evidence of water in the micellar core.

We consider now the small-angle X-ray scattering experiments of Svens and Rosenholm [36] and Friman and Rosenholm [37]. In both studies, the micelles are modelled as two regions – a dry spherical hydrophobic core with the electron density of pure hydrocarbon ($-73 e^-/nm^3$ *with respect to the solvent*) with a radius R_{par} surrounded by a spherical polar annulus extending from R_{par} to R_{pol} and with an electron density of ϱ_{pol}.

In both studies, the X-ray scattering curve is fitted assuming *no* intermicellar spatial correlation and using four adjustable parameters: R_{par}, R_{pol}, ϱ_{pol} and a scaling constant for the whole scattering curve. Here, we discuss one of the systems studied (1.0 M sodium octanoate with no added salt) as it is indicative of all the other results.

Svens and Rosenholm get a best fit to their scattering curve with the values $R_{par} = 4.5$ Å, $R_{pol} = 26.5$ Å and $\varrho_{pol} = 0.4\,e^-/nm^3$ (with respect to the solvent) while Friman and Rosenholm find $R_{par} = 5.6$ Å, $R_{pol} = 24.3$ Å and $\varrho_{pol} = 1.4\,e^-/nm^3$ (with respect to the solvent). Such model micelles are very hard to comprehend. About 99% by volume of these micelles (the polar region for which $R_{par} < r < R_{pol}$) is deduced to have an electron density hardly different from the surrounding solvent and hence is almost invisible to the X-rays being used to probe its structure. Further, the fully extended length of an octanoate molecule is ~13 Å. So, amphiphiles with their headgroup in the outer regions of the polar annulus of these model micelles have *no part of their chain in the micelle's dry core.*

As Friman and Rosenholm are aware, there is substantial interparticle spatial correlation at the solution conditions studied by them. Their measurements extend out to scattering vectors s ($= 2\sin\theta/\lambda$) of $0.5\,nm^{-1}$ equivalent to a momentum transfer Q ($= 4\pi\sin\theta/\lambda$) of $3.1\,nm^{-1}$. As shown by Hayter and Zemb [43] for the same system, the structure factor describing intermicellar spatial correlations, $S(Q)$, shows *considerable structure* out to this value of Q (see figure 2, Hayter and Zemb).

In fitting their neutron scattering data, Hayter and Zemb *assume* a dry spherical core containing all but one CH_2 group per amphiphile chain surrounded by a spherical polar annulus. Including $S(Q)$ in their analysis, they fit their experimental scattering curve with two parameters, the mean aggregation number of the micelles and the micellar net charge. Under experimental conditions almost identical to the X-ray studies (1.05 M sodium octanoate with no added salt), the Hayter and Zemb fitted parameters imply (using the above notation) $R_{par} = 9.8$ Å and $R_{pol} = 13.2$ Å. The fit they obtain to their experimental scattering curve using two free parameters *is substantially better* than the fits obtained to the X-ray scattering data using four free parameters (compare figure 2 in [43] with figure 3 in [36] or with figure 1 in [37]).

Without necessarily committing oneself to the assumptions made by Hayter and Zemb, one must conclude that the values deduced for R_{par} and R_{pol} in [36] and [37] are very unreliable. They do not provide grounds for challenging the standard picture of ionic micelles.

We turn now to three arguments presented by Menger and Doll [34] in support of the Menger micelle.

(i) They discuss the consequences of an observation by Zana that in an annulus of about 4 Å width out from the micelle's hydrophobic core where headgroups, chains and water mix, there is a region where the environment would not be too different from that of bulk water. Menger and Doll comment: "Assume for the moment that this statement is correct. Simple volume calculations would then demand that fully 69% of an SDS micelle (*not* counting head groups) is water-like!". A very similar comment applies to the model micelles analysed in section 2. For about 4 Å out from the hydrophobic core, chains and headgroups do mix and the environment is not too different from bulk water. Regarding the micelle as a sphere with a radius 4 Å larger than the core radius, a large fraction (by volume) of the micelle is indeed water-like.

So the Menger and Doll statement is irrefutable but also misleading. The model amphiphile chains have segregated themselves away from the water and their own headgroups *almost as much as is physically possible.* 94% of the chain volume resides inside the dry core and the hydrocarbon volume fraction falls off rapidly outside the core (fig. 8). There *is* contact of the chains with water, simply because of the large area per amphiphile at the micelle surface (more than 56 Å^2 for the micelle of figure 8).

(ii) Menger and Doll quote Almgren and Swarup [44] in support of the statement: "Micellar radii are known to *exceed* the length of a fully extended chain plus headgroup." Menger and Doll continue: "This curious observation can be accommodated by a loosely packed and disordered micelle in possession of rough surfaces. Water fills the irregularities, ...". In fact, Almgren and Swarup [44] measure aggregation numbers rather than micellar radii. Their aggregation numbers all seem consistent with a dry hydrophobic core which is globular, not too aspherical, and which contains almost all the amphiphile chains.

As we have seen in section 2, the results of Bendedouch et al. [11] are best fitted assuming a dry hydrophobic core with a radius slightly larger than the fully extended chain length. The same is true of a recent extensive neutron diffraction study of tetradecyl trimethylammonium bromide micelles [45]. Twenty different scattering curves (obtained by varying the scattering length density of the solvent) can be well fitted assuming dry spherical micelle cores with a mean radius of 22 Å. Since the fully extended length of these chains is ~19.3 Å, a perfectly spherical core has a hole of ~82 Å^3 at its centre. As discussed in section 2, small deviations from a perfectly spherical core would eliminate this hole. Again, the spherical shape probably makes a non-negligible contribution to the ensemble of equilibrium shapes of these micelles.

(iii) Menger and Doll study the rate of oxidation of double bonds in the chains of micellized amphiphiles as a function of position of the double bond. The oxidation takes place when the double bond is either in the aqueous phase or in close proximity to it. They demonstrate that a terminal double bond is significantly more exposed to water than double bonds nearer the centre of the chain.

It seems likely that the double bond shows a slight preference for the micelle interface over the micelle interior due to its small associated polarity. Ignoring this effect, we may compare results of this experiment with the single chain model presented in section 2. Figure 7 demonstrates that *all* chain segments spend some time in contact with the aqueous phase and that the terminal CH_3 group is indeed in contact with the surface a larger fraction of time than segments nearer the chain centre. As we have already remarked in discussion of Figure 9, this occurs partly because many chains venture some way into the core before returning to the surface. The single chain model therefore provides a satisfactory explanation for these experiments. The standard picture, on which the single chain model is based, is quite secure.

4. Conclusion

1. Based on a wide range of experimental and theoretical evidence, a standard picture of the state of the amphiphiles in ionic micelles is presented.

2. Using this standard picture as a starting point, a simple quantitative model for the amphiphile chains is presented. The model is based on generating all conformations of a single chain. As is demonstrated, over a wide range of static properties, the single chain model provides close agreement with the results of a large molecular dynamics simulation which explicitly considers the interactions of 128 amphiphiles in a bilayer.

3. The single chain model is used to generate results for spherical micelles. It predicts the following things which are in good agreement with experiment:

(i) the order parameter profile down the amphiphile chain,

(ii) the mean and standard deviation of position of the terminal methyl group,

(iii) that all segments of the chain spend some time in contact with the aqueous phase and that the terminal group is exposed to the polar exterior more often than groups nearer the chain centre,

(iv) that on average almost all the chain volume is in a dry hydrophobic core,

(v) that spherical micelles with a core radius equal to or slightly greater than the fully extended chain length are formed at only small free energy cost to the chains. In forming such micelles, the chains do not lose their conformationally disordered (liquid-like) nature.

4. Evidence which has been claimed to invalidate the standard picture of ionic micelles is examined in detail and found wanting.

Acknowledgements

I am most grateful to Bertil Halle, Jacob Israelachvili and Fred Menger for their helpful comments on the manuscript. I also thank Fred Menger for sending me a copy of reference [38].

References

1. Flory PJ (1969) Statistical Mechanics of Chain Molecules. Wiley-Interscience, New York
2. Wennerström H, Lindman B, Söderman O, Drakenberg T, Rosenholm JB (1979) J Am Chem Soc 101:6860
3. Walderhaug H, Söderman O, Stilbs P (1984) J Phys Chem 88:1655
4. Schatzberg P (1963) J Phys Chem 67:776
5. Büldt G, Gally HU, Seelig A, Seelig J, Zaccai G (1978) Nature Lond 271:182
6. Dilger JP, Fisher LR, Haydon DA (1982) Chem Phys Lipids 30:159
7. Wennerström H, Lindman B (1979) J Phys Chem 83:2931
8. Tanford C (1980) The Hydrophobic Effect: Formation of Micelles and Biological Membranes, 2nd ed. Wiley-Interscience, New York
9. Halle B, Carlström G (1981) J Phys Chem 85:2142
10. Cabane B (1981) J Physique 42:847
11. Bendedouch D, Chen S-H, Koehler WC (1983) J Phys Chem 87:153
12. Cabane B, Duplessix R, Zemb T (1984) In: Surfactants in Solution, Vol 1, Mittal KL, Lindman B (eds) Plenum Press, New York, p 373 – 404
13. Vikingstad E, Høiland H (1978) J Colloid Interface Sci 64:510
14. Kennedy GC, Keeler RN (1972) American Institute of Physics Handbook, Section 4d: Compressibility. McGraw-Hill Co
15. Dill KA, Flory PJ (1981) Proc Natl Acad Sci USA 78:676
16. Dill KA (1982) J Phys Chem 86:1498
17. Hartley GS (1936) Aqueous Solutions of Paraffin-Chain Salts: A Study in Micelle Formation. Hermann and Co, Paris
18. Fromherz P (1981) Ber Bunsenges Phys Chem 85:891
19. Gruen DWR (1981) J Colloid Interface Sci 84:281
20. Gruen DWR, de Lacey EHB (1984) In: Surfactants in Solution, Vol 1, Mittal KL, Lindman B (eds). Plenum Press, New York, p 279 – 306
21. Gruen DWR (1985) J Phys Chem 89:146, 153
22. Marcelja S (1974) Biochem Biophys Acta 367:165
23. van der Ploeg P, Berendsen HJC (1983) Mol Phys 49:233

24. Seelig J, Niederberger W (1974) Biochemistry 13:1585
25. Gruen DWR (1980) Biochim Biophys Acta 595:161
26. Flory PJ (1979) Faraday Discuss Chem Soc 68:14
27. Yoon DY, Flory PJ (1978) J Chem Phys 69:2536
28. Vacatello M, Avitabile G, Corradini P, Tuzi A (1980) J Chem Phys 73:548
29. Jorgensen WL (1982) J Chem Phys 77:5757
30. Jorgenson WL (1983) J Phys Chem 87:5304
31. Weber TA (1979) J Chem Phys 70:4277
32. Aniansson GEA (1978) J Phys Chem 82:2805
33. Menger FM (1979) Acc Chem Res 12:111
34. Menger FM, Doll DW (1984) J Am Chem Soc 106:1109
35. de Mayo P, Sydnes LK (1980) J C S Chem Comm 994
36. Svens B, Rosenholm JB (1973) J Coll Interface Sci 44:495
37. Friman R, Rosenholm JB (1982) Colloid Polym Sci 260:545
38. Whitten DG, Russell JC, Foreman TK, Schmehl RH, Bonilha J, Braun AM, Sobol W (1982) In: Chemical Approaches to Understanding Enzyme Catalysis: Biomimetic Chemistry and Transition-State Analogs. Green BS, Ashani Y, Chipman D (eds). Elsevier, Amsterdam
39. Zachariasse KA, Van Phuc N, Konzankiewicz B (1981) J Phys Chem 85:2676
40. Mukerjee P, Cardinal JR, Desai NR (1977) In: Micellization, solubilization and microemulsions. Mittal KL (ed). Plenum Press, New York 1:241
41. Lindman B, Wennerström H, Gustavsson H, Kamenka N, Brun B (1980) Pure Appl Chem 52:1307
42. Ganesh KN, Mitra P, Balasubramanian D (1984) In: Surfactants in Solution, vol 1. Mittal KL, Lindman B (eds). Plenum Press, New York, p 599 – 611
43. Hayter JB, Zemb T (1982) Chem Phys Lett 93:91
44. Almgren M, Swarup S (1983) J Coll Interface Sci 91:256
45. Tabony J (1984) Mol Phys 51:975

Received June 6, 1984;
accepted October 15, 1984

Author's address:

Dr. David W. R. Gruen
Dept. of Applied Mathematics
Research School of Physical Sciences
Australian National University
Canberra, A.C.T. 2601 (Australia)

Progress in Colloid & Polymer Science　　　　　　　Progr Colloid & Polymer Sci 70:17 – 22 (1985)

Phase equilibria in systems containing both an anionic and a cationic amphiphile. A thermodynamic model calculation

P. Jokela, B. Jönsson, and H. Wennerström

Division of Physical Chemistry 1, Chemical Center, Lund (Sweden)

Abstract: We have studied the equilibrium between a lamellar phase and crystalline salts in a four-component system containing both a cationic and an anionic surfactant in water. The chemical potentials for the different components have been derived from a thermodynamic model for the free energy.

In our model system there are four salts that can precipitate: A^+X^-, A^-X^+, A^+A^- and X^+X^-. A^+ and A^- are, respectively, a cationic and an anionic surfactant; X^- and X^+ are the corresponding counterions. The system is represented as a trigonal bipyramid. We have calculated the equilibrium between a lamellar phase and crystalline phases A^+A^-, A^-X^+ and A^+X^- with different weight fractions of A^+A^-.

The observed instability gap between the two lamellar regions around equal amounts of A^+X^- and A^-X^+ can be explained by the fact that dispersion attraction between the amphiphilic plates dominates over the electrostatic repulsion. The high entropy in the lamellar liquid crystal compared with a common crystal can be the reason for the extension of the lamellar stability region to very low water contents in the vicinity of the instability gap.

Key words: Phase equilibria, thermodynamic model, four-component systems

Introduction

Surfactant-water systems show a rich polymorphism and in addition to isotropic solutions and crystals a range of different liquid crystalline phases can form [1, 2]. The phase behaviour provides an essential clue to the understanding of the properties of surfactant systems and a large amount of experimental work has been devoted to determine phase equilibria. The theoretical description of the thermodynamic properties of the system has been studied to a much lesser extent.

In a series of papers we have developed a model for the thermodynamic properties of ionic surfactant-water systems [3 – 5]. This has led to fairly satisfactory description of phase equilibria in two and three component systems. As a continuation of this work we present a study of a model four-component system $H_2O - A^-X^+ - A^+X^-$ where A^- and A^+ are an anionic and a cationic surfactant, respectively, while X^+X^- forms an ordinary electrolyte. Dealing with four components in studies of phase equilibria poses a substantial problem due to the complexity

introduced by having at least three degrees of freedom within the one-phase areas. A model study is helpful in sorting out some of these difficulties.

The representation of the four-component system

A system produced by a solvent (water) and two simple salts A^+X^- and A^-X^+ with no common ion has four components in the sense used in connection with the Gibbs phase rule. There are four possible simple crystalline salts for the system since in addition to A^+X^- and A^-X^+ one can have A^+A^- and X^+X^-. In principle all four of these salts can form and to have a useful graphical representation of the system all of them should be represented as a point in the diagram. For given values of the intensive variables temperature and pressure the conventional triangular diagram for a three component system can be extended using a trigonal bipyramid as illustrated in figure 1. In this diagram there are three degrees of freedom and every possible composition can be represented.

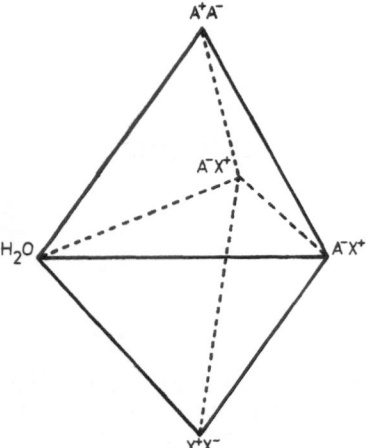

Fig. 1. A trigonal bipyramid representing the four-component system $H_2O - A^-X^+ - A^+X^- - A^+A^-$

The model system

We have chosen to study a model system containing an anionic surfactant A^-X^+ and a cationic surfactant A^+X^- where X^+ and X^- are assumed to be simple ions as Na^+ and Cl^-, respectively. Such a system could show interesting and unique properties, which should be sensitive to the ratio A^+/A^-. A particular feature is that the crystal A^+A^- is expected to be particularly stable. As a consequence we have concentrated on the possible equilibria with such a crystalline phase. For most soaps it is the lamellar phase that is the most stable liquid crystalline phase at higher concentrations and this is the other phase we consider. Thus the aim is to describe the equilibrium between crystalline phases A^+A^-, A^-X^+ and A^+X^- and a lamellar phase with varying ratio A^+/A^-. The simple salt X^+X^- is considered strongly water soluble. As a consequence of the restriction to one

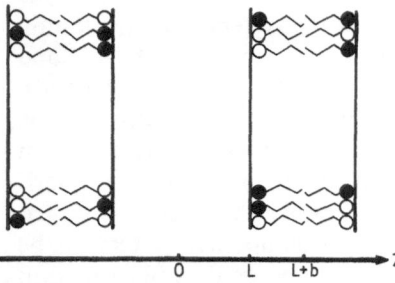

Fig. 2. Schematic representation of a lamellar system. $2L$ is the thickness of water layer and $2b$ the thickness of surfactant bilayer

liquid crystalline phase only a partial phase diagram is obtained. However, since our primary aim is to illustrate some general properties of a particular type of four-component system the restriction to few phases only provides a useful simplification.

The lamellar phase consists of aqueous layers with thickness, $2L$, alternating with surfactant bilayers, thickness $2b$, as illustrated in figure 2. The surfactant molecules are assumed to be totally incorporated into the lamella making the average charge density determined by the composition and the area per molecule. The ions X^+ and X^- are assumed to reside in the aqueous layer, where they partly have the role of counterions, X^+ or X^- depending on the A^+/A^- ratio, and partly the role of a screening electrolyte.

The thermodynamic model

Most parts of the thermodynamic model used have been presented previously, in [3 – 5]. We will therefore here only give a brief summary of the different energy contributions and their effect on the chemical potentials.

The free energy G of a lamellar system according to the model is written as

$$G = n_{H_2O}\mu^0_{H_2O} + n_{A^-X^+}\mu^0_{A^-X^+} + n_{A^+X^-}\mu^0_{A^+X^-}$$
$$+ n_{A^-A^+}\mu^0_{A^-A^+} + G_S + G_{el} - TS_{mix} + G_{DISP}$$

$$(1)$$

where μ^0_i is the standard chemical potential of component i, G_S a surface free energy, G_{el} an electrostatic free energy, S_{mix} the entropy from the mixing of the components and G_{DISP} the free energy from the dispersion interactions.

G_S the surface free energy is assumed to follow the empirical equation [5]

$$G_S = \gamma_+ \cdot n_{A^+} \cdot A_+ + \gamma_- \cdot n_{A^-} \cdot A_-$$

$$(2)$$

where γ_+ and γ_- are experimentally determined constants, A_+ and A_- the surface area of the cationic respectively anionic amphiphile. The total area A between water and hydrocarbon parts may then be written as

$$A = n_{A_+}A_+ + n_{A_-}A_-.$$

$$(3)$$

The contributions to the chemical potentials from the surface free energy will then be

$$\mu_{A^+}^S = \gamma_+ A_+ \qquad (4a)$$

$$\mu_{A^-}^S = \gamma_- A_- . \qquad (4b)$$

The contribution to the chemical potentials from the electrostatic free energy G_{el} and the entropy of mixing TS_{mix} are [3]

$$\mu_{H_2O}^{el\text{-}TS} = -N_A kT V_{H_2O} \cdot (c_+(0) + c_-(0)) \qquad (5a)$$

$$\mu_{A^-X^+}^{el\text{-}TS} = kT \ln(c_+(L) \cdot X_-) - A_- \cdot 2E_{el}/A \qquad (5b)$$

$$\mu_{A^+X^-}^{el\text{-}TS} = kT \ln(c_-(L) \cdot X_+) - A_+ \cdot 2E_{el}/A \qquad (5c)$$

$$\mu_{X^+X^-}^{el\text{-}TS} = kT \ln(c_+(0) \cdot c_-(0)) \qquad (5d)$$

for a lamellar system. Here V_{H_2O} is the volume of a water molecule, $c_i(0)$ and $c_i(L)$ are the concentrations of component i at $z = 0$ respectively $z = L$, (cf. fig. 2). X_i is the mole fraction in the amphiphilic aggregate of component i and E_{el} the energy of the ion-ion interaction. This is for a lamellar phase [4].

$$E_{el} = kT \cdot \{(n_{A^-X^+} + n_{A^+X^-} + 2 \cdot n_{X^+X^-}) - N_A \cdot L$$
$$\cdot A \cdot (c_+(0) + c_-(0))\} . \qquad (6)$$

The dispersion energy G_{DISP} may for an infinite array of water and hydrocarbon sheets be written as [6]

$$G_{DISP}/A = -\frac{H}{48\pi b^2} \left\{ \frac{1}{2} + \sum_{j=1}^{\infty} \frac{y^4}{j^4} \cdot \frac{3 - (y^2/j^2)}{[1 - (y^2/j^2)]^2} \right\}$$
$$= -f(y)/b^2 \qquad (7)$$

where H is the Hamaker constant, $2b$ the thickness of a hydrocarbon sheet and y the volume fraction of hydrocarbon

$$y = \frac{b}{b+L} . \qquad (8)$$

The dispersion energy is in equation (7) assumed to be zero when $L \to \infty$ and $b \to \infty$. The contributions to the chemical potentials from the dispersion energy will then be

$$\mu_{H_2O}^{DISP} = \frac{V_{H_2O}}{b^3} \cdot y^2 \cdot f'(y) \qquad (9a)$$

$$\mu_{A^-X^+}^{DISP} = -\frac{A_-}{b^2}(f(y) + (y - y^2) \cdot f'(y)) \qquad (9b)$$

$$\mu_{A^+X^-}^{DISP} = -\frac{A_+}{b^2}(f(y) + (y - y^2) \cdot f'(y)) . \qquad (9c)$$

The chemical potentials may now be obtained by combining equations (1), (4), (5) and (9).

$$\mu_{H_2O} = \mu_{H_2O}^0 - N_A kT V_{H_2O}(c_+(0) + c_-(0)) +$$
$$+ \frac{V_{H_2O}}{b^3} \cdot y^2 \cdot f'(y) \qquad (10a)$$

$$\mu_{A^-X^+} = \mu_{A^-X^+}^0 + kT \ln(c_+(L) \cdot X_-) + A_-$$
$$\cdot \{\gamma_- - 2E_{el}/A - (f(y) + (y - y^2)f'(y)/b^2\} \qquad (10b)$$

$$\mu_{A^+X^-} = \mu_{A^+X^-}^0 + kT \ln(c_-(L) \cdot X_+) + A_+$$
$$\cdot \{\gamma_+ - 2E_{el}/A - (f(y) + (y - y^2)f'(y)/b^2\} \qquad (10c)$$

$$\mu_{X^+X^-} = \mu_{X^+X^-}^0 + kT \ln(c_+(0) \cdot c_-(0)) \qquad (10d)$$

$$\mu_{A^+A^-} = \mu_{A^+X^-} + \mu_{A^-X^+} - \mu_{X^+X^-} . \qquad (10e)$$

In applying the equations for the chemical potentials one must know the areas A_+ and A_- of the amphiphilic molecules as well as the thickness of the hydrocarbon sheet $2b$. These quantities may for systems with low charge densities be calculated from the volume V_i and the maximum length $b_{i\,max}$ of the amphiphile molecules.

$$A_i = V_i/b_{i\,max} . \qquad (11)$$

The thickness of the hydrocarbon sheet will then be

$$2b = 2 \cdot \frac{n_{A_+} \cdot V_{A_+} + n_{A_-} \cdot V_{A_-}}{n_{A_+} \cdot A_{A_+} + n_{A_-} \cdot A_{A_-}} . \qquad (12)$$

This procedure may however not be used for systems with high charge densities since equation (11) is then no longer valid (see reference 4). The condition for optimal size

$$\left(\frac{\partial G}{\partial b}\right)_{n_j} = 0 \qquad (13)$$

must instead be used to determine A_i and b.

The partial derivative $(\partial G/\partial b)_{n_j}$ is

$$\left(\frac{\partial G}{\partial b}\right)_{n_j} = -A \cdot \{\gamma - 2E_{el}/A - 3 \cdot f(y)/b^2\}/b \qquad (14)$$

following Appendix 2 of reference 4.

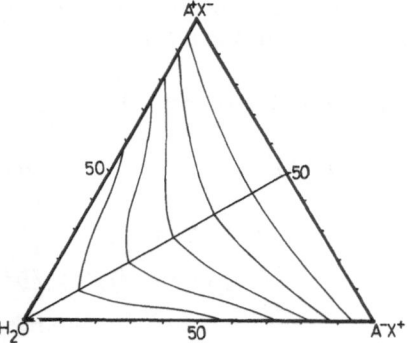

Fig. 3. The chemical potential of water remains unchanged along the curves drawn in the phase diagram

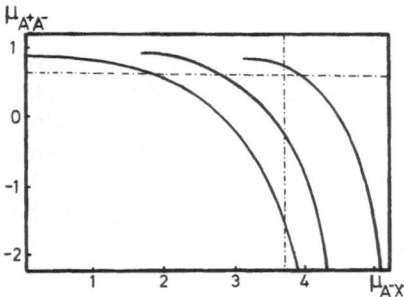

Fig. 4. The full lines represent the chemical potential of A^+A^- ($\mu_{A^+A^-}$) as a function of the chemical potential of A^-X^+ ($\mu_{A^-X^+}$) for three different L-values. The dotted lines show the constant chemical potentials of the crystalline salts

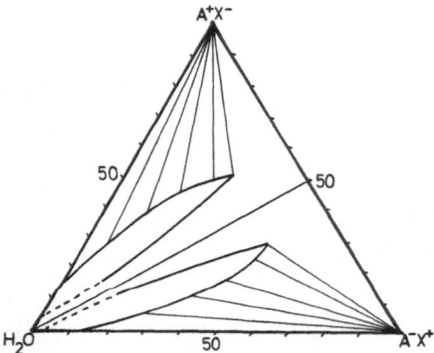

Fig. 5. Tie-lines between a lamellar phase and the crystalline phase A^+X^- and A^-X^+

The computational procedure

For two or more phases in equilibrium the chemical potentials of all components should be equal in the two phases. Since the expressions for the chemical potentials in equations (10a − e) are fairly complex it poses a considerable computational problem to solve for the phase boundaries. At the present stage we are only considering one liquid crystalline phase, which introduces a significant simplification of the calculations. For the crystalline phases fixed values are given for the crystallized component and for the other components the chemical potentials vary very strongly with minute additions of the particular component. Consequently one can satisfy the equilibrium with the liquid crystal for any values of the chemical potential of the impurity. It is assumed that mixed crystals do not form.

For the lamellar phase equations (10a − e) apply. This phase can in the model be in equilibrium with either of the three crystals A^+X^-, A^-X^+ and A^-A^+. To find these equilibria a stepwise procedure is used.

i) Confine the composition in the lamellar phase to a particular plane $H_2O - A^-X^+ - A^+X^-$ in the bipyramid.

ii) Choose a value for the water chemical potential.

iii) Choose a value for the thickness, $2L$, of the aqueous layer.

iv) Choose a value for $\alpha = c_+(0)/c_-(0)$. Using equation (10a) and the values assigned in ii) − iv) $c_+(0)$ and $c_-(0)$ are calculated.

v) Choose a value for the thickness, $2b$, of the surfactant bilayer.

vi) Solve the *PB* equation (see reference 7) and check that the value assigned to b is consistent with equations (12) and (14). If not go back to iv).

vii) Determine the composition from the solution of the *PB* equation. If it is not consistent with the particular plane in the bipyramid chosen in i) go back to v).

viii) Calculate $\mu_{A^-A^+}$ and $\mu_{A^-X^+}$. Go back to iii).

ix) When $\mu_{A^-A^+}$ and $\mu_{A^-X^+}$ have been calculated for a number of different L-values they are plotted in a diagram as shown in figure 3 and figure 4.

x) The chemical potentials of the crystalline salts are assigned particular values. They represent straight lines in figure 4. Equilibrium between a crystalline phase and the lamellar phase is obtained at the intersections in the plot.

xi) The composition in the lamellar phase for the two intersection points is calculated and one obtains one tie-line to a crystalline phase as illustrated for A^-X^+ and A^+X^- crystals in figure 5.

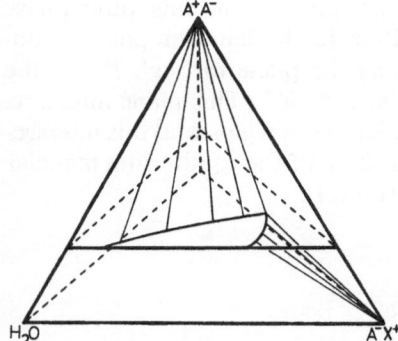

Fig. 6. The extension of the lamellar phase for a particular plane of the bipyramid and the tie-lines between the lamellar phase and the crystalline phases A^-X^+ and A^+A^-

xii) The procedure is now repeated for a new choice of the water chemical potential.

xiii) The full phase diagram is obtained by choosing a new plane in the bipyramid.

Results

Calculations were performed for compositions in the upper part of the bipyramid since this is the region where equilibria with A^+X^-, A^-X^+ and A^+A^- crystals are of most interest. There is furthermore complete symmetry with respect to charge inversion in the model so it is sufficient to consider only one half of the bipyramid.

Figure 6 shows for a particular plane of the bipyramid the extension of the lamellar phase and also the tie-lines. Figures 7a – d show the extension of the lamellar phase for a series of planes.

Discussion

The problems of applying the previously developed thermodynamic model to a four-component system has been investigated. It is shown that at least for the restricted applications to equilibria between a lamellar phase and crystals the computational problems can be solved using only a minicomputor. We further expect that the calculations will be of considerable help in the experimental studies of anionic surfactant-cationic surfactant-water systems that are currently performed in this laboratory.

Due to the conceptual simplicity of the model it is fairly straightforward to interpret the resulting phase

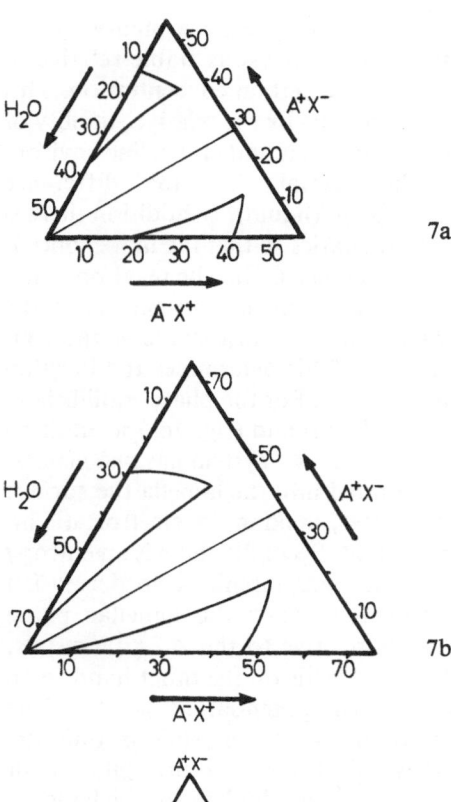

7a

7b

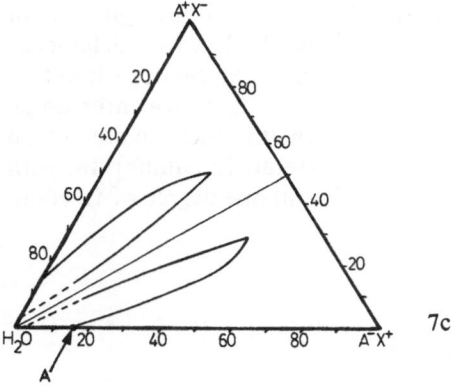

7c

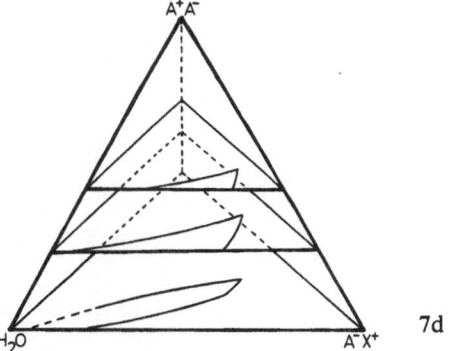

7d

Fig. 7. The extension of the lamellar phase for a series of different weight fractions of A^+A^-. a) 45% w/w, b) 25% w/w and c) 0% w/w. d) The three planes of a) – c) represented in the bipyramid

equilibria qualitatively. For the pure binary system $H_2O - A^-X^+$ the lamellar phase is stable relative to the crystalline salt up to a certain concentration. This concentration is determined by the relative stability of the surfactant molecule in a liquid crystalline environment and in the crystal, i.e. the difference $\mu^0_{A^-X^+}$ (crystal) $- \mu^0_{A^-X^+}$ (liquid). In addition there is an effect of the electrostatics so that the more concentrated the system the higher is the chemical potential of the surfactant in the lamellar system. Thus the crystal will ultimately be the most stable as the concentration is increased. This determines the location of the point A in figure 7c. For the phase equilibria in the base plane of the bipyramid (fig. 7c) the addition of A^+X^- to the $H_2O - A^-X^+$ system has two effects. Since A^+ is incorporated into the lamella the surface charge density decreases, making the electrostatic interactions less important. Secondly there is an entropy of mixing term for the amphiphiles also decreasing their chemical potential. Thus the lamellar phase becomes more stable relative to the A^-X^+ crystals. The addition of A^+X^- leads on the other hand to an increase in the chemical potential of A^+A^-. The stability relative to the A^+A^- crystals is thus decreased. Eventually this leads to a precipitation of these crystals giving a new border line for the lamellar phase region. The point where the two precipitation limits meet lies on one of the edges of a three-phase region in the bipyramid. In this three-phase region with A^-X^+ and A^+A^- crystals in equilibrium with the lamellar phase there is still one degree of freedom left. For an arbitrary point P in this three-phase region the composition in the lamellar phase is obtained by constructing the plane through P and the two corners A^+A^- and A^-X^+. This plane intersects the edge of the lamellar phase region and this intersection gives the composition of the equilibrium lamellar phase at the composition P.

References

1. Ekwall P (1975) In: Brown G (ed) Advances in Liquid Crystals, Vol 1. Academic Press, New York, p 1
2. Tiddy GJT (1980) Phys Rep 37:1
3. Jönsson B, Wennerström H (1981) J Colloid Interface Sci 80:482
4. Jönsson B (1981) Thesis, Lund
5. Jönsson B, Nilsson P-G, Lindman B, Guldbrand L, Wennerström H (1984) In: Mittal KL, Lindman B (eds) Surfactants in Solution, Vol 1. Pergamon Press, New York, p 1
6. Mahanty J, Ninham BW (1976) Dispersion Forces. Academic Press, London New York San Fransisco
7. Jönsson B, Wennerström H (1980) Chemica Scripta 15:40

Received July 30, 1984;
accepted November 20, 1984

Authors' address:

Päivi Jokela
Div. Phys. Chem. 1
Chemical Center
P.O. Box 124
S-22100 Lund (Sweden)

Progress in Colloid & Polymer Science   Progr Colloid & Polymer Sci 70:23 – 29 (1985)

The size, shape and hydration of micelles in aqueous medium

K. S. Birdi

Fysisk-Kemisk Institut, The Technical University of Denmark, Lyngby (Denmark)

Abstract: The size, shape and hydration of micelles (Triton-X-100, NaDDS, NP-10, NP-13, NP-18) in aqueous media were estimated based on micellar molecular weight and intrinsic viscosity data. The hydration values were estimated by using the Simha factor. The most consistent data (i.e. degree of hydration) were obtained when assuming an ellipsoidal shape of Triton-X-100 micelles as being oblate, while prolate shapes did not give useful values, both as regards hydration and axial ratios. However, the analyses of other micellar systems did not provide a definite conclusion as regards the ellipsoidal shape.

Key words: Micelle shape, micelle hydration

List of abbreviations

N aggregation number
N_n number average aggregation number
N_w weight average aggregation number
R_{Stokes} Stokes radius
R_{hc} radius of hydrocarbon core of micelle
L_{hc} length of hydrocarbon chain
L_p length of polar groups
V^m volume of micelle
V^l_{hc} volume of hydrocarbon chain
V^m_{hc} volume of hydrocarbon core
V^m_p volume of polar group per chain
pr   prolate
ob   oblate
V^m_{Stokes} volume of micelle from R_s

Introduction

Molecules containing both hydrophobic and hydrophilic parts, amphiphile molecules, under the influence of the hydrophobic interactions in aqueous media form aggregates, micelles, when the concentration is greater than the critical micelle concentration (CMC). In this aggregation process, the hydrocarbon chains are located inside the micelle core such that this behaves as "liquid-like" alkane, while the hydrophilic groups maintain contact with the continuous solvent phase (water) at the micelle-water interface. In order to be able to describe the standard free energy of formation of micelles, many theories have been proposed by various investigators in the current literature.

In the same context the determination of the size and shape of micelles in aqueous media is a subject of considerable current interest and much controversy [1 – 22]. These theoretical treatments have indicated that the size, shape and hydration of micelles, in-aqueous media, are of great importance. In one of these reports [9] the calculation of the size, shape and hydration of micelles formed by a nonionic amphiphile, Triton-X-100, were performed based on certain assumptions, as regards the density of the hydrocarbon core of the micelle and the relation of this volume of the core to the length of the alkyl chain of amphiphile, as estimated from the molecular models. From these analyses it was concluded that the micelle aggregates of Triton-X-100, based on an *average* value of different data from reported literature values, did not correspond to a spherical shape for the hydrocarbon care [9]. Intrinsic viscosity is a function of the size and shape of the particle and solvent-solute interactions. Further, from the intrinsic viscosity data and the Simha viscosity factor, it was found that reasonable hydration values were obtained only when it was assumed that the shapes of both the hydrocarbon core and the composite micelle (i.e. hydrocarbon core plus the hydrophilic groups with the entrapped water) were of oblate ellipsoid rather than prolate ellipsoid. This micelle was also analyzed by using simple geometrical packing considerations, but the analysis did not give any specific conclusion as regards the oblate or

Table 1. Aggregation number (N), hydrocarbon core volume (V_{hc}^m), radius of core and L_{hc} for ionic and nonionic (Triton x-100) micelles

Surfactant	H$_2$O	N	V_{hc}^m (Å^s)	R_{hc} (Å)	L_{hc} (Å)	(Ref.)
n-C$_8$H$_{17}$SO$_4$Na	H$_2$O	31	6710	11.7	11.6	(H)
n-C$_{10}$H$_{21}$SO$_4$Na	H$_2$O	38	10264	13.5	14.2	(H)
n-C$_{11}$H$_{23}$SO$_4$Na	H$_2$O	45	13360	14.7	15.4	(H)
n-C$_{12}$H$_{25}$SO$_4$Na (NaDDS)	H$_2$O	60	19380	17	16.7	(H)
n-C$_{12}$H$_{25}$SO$_4$Na (NaDDS)	0.6 M NaCl	183	59100	24.5	16.7	(B)
Triton-x-100	H$_2$O	143	51900	23	10	(Ro)

$$\left[V_{hc}^m = \frac{N(\text{mol} \cdot \text{wt of alkyl part}) (10^{24}) (\text{Å}^3)}{0.87 \, (\text{g/cm}^3) \, 6.02 \cdot 10^{23} \, (\text{Avogadros number})} \, ; \, R_{hc} = \left(\frac{3 V_{hc}}{4 \Pi} \right)^{1/3} ; \, L_{hc} = (1.5 + 1.265 \, n_c) \right]$$

[(H) Huisman (1964); (B) Birdi et al. (1980); (Ro) Robson & Dennis (1977)]

prolate ellipsoid as the preferred shape, as one would have expected (to be described herein).

Since the analysis of size and shape of micelles constitutes a rather important basis of knowledge, the example of merely one set of data, i.e. Triton-X-100, does not provide enough information to make any general conclusions. Further, the data used as basis for these conclusions were some average value for intrinsic viscosity 5.3 cm^3/g (4.7 – 5.5 range) and aggregation number, $N = 143$ (100 – 143 range). The present study is an attempt to supply this lacking information by providing more systematic data, with the adequate analysis, as regards size, shape and hydration of various micelles, for both nonionic and ionic micelles.

Theoretical

Spherical or ellipsoidal micellar shapes: In the consideration of micellar shapes, e.g. either ionic or nonionic, we can for the sake of simplification analyze the differences in the free energy of forming a spherical or ellipsoidal micellar shape. We will consider spherical micelles as being those which are not restricted as regards packing constraints to other possible shapes (ellipsoids, cylinders or bilayers), and thus would constitute the smallest aggregate with a given surface area per micelle, wherefrom it would be the aggregate with the lowest aggregation number, at a given temperature and pressure. It can be argued that a spherical micellar shape would constitute a structure with the lowest surface free energy, and would thus be thermodynamically the most favoured packing structure, due to the absence of varying curvature effects as one would find in ellipsoids. It is obvious that micelles formed by nonlinear surfactants, e.g. bile salts etc., cannot be analyzed by these theo-

ries, due to the fact that steric hindrance gives rise to rather small aggregation numbers, 10 – 20 [23]. For a given aggregation number, a spherical micelle with the radius R_s and total volume V_s, with surface area per micelle A_s, we would have:

$$V_s = \tfrac{4}{3} \Pi R_s^3 \quad \text{and} \quad A_s = 4 \Pi R_s^2$$
$$R_s = 3 \cdot V_s / A_s \tag{1}$$
$$\simeq L_{\max} .$$

Thus, if the fully stretched amphiphile molecule is taken as $L_{\max}$, then in the case of a spherical micelle $R_s \simeq L_{\max}$. Further, any deviation from the relation between R_s and $L_{\max}$ as given in equation (1) would merely indicate that the micelle is ellipsoidal.

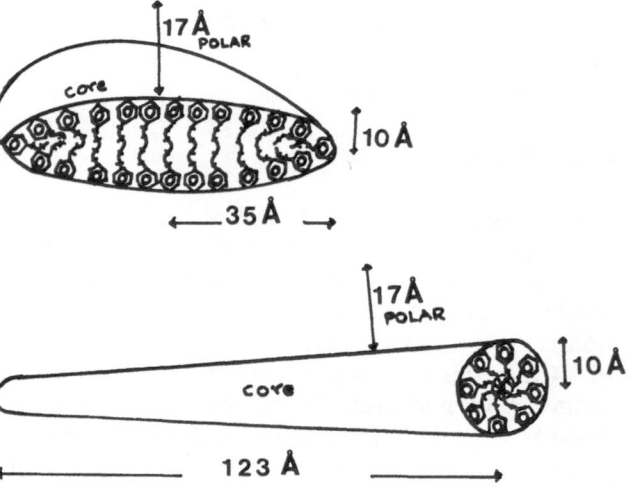

Fig. 1. The hydrocarbon core and the whole micelle with ellipsoidal shapes: oblate or prolate (cutaway views), for Triton-X-100

This simple approach [15] becomes complicated when considering real systems, since one generally describes the hydrocarbon core and the polar part of the micelle separately. We will pursue the analysis accordingly, as follows.

Hydrocarbon core: The approach which has been generally used has been to assume that the hydrocarbon core of a micelle is equivalent to a "liquid-like hydrocarbon", which has a density of approximately that of liquid hydrocarbon [15]. If one knows the aggregation number, then one can calculate the volume of the hydrocarbon core. Then knowing the maximum length for the alkyl chain when fully stretched, L_{hc}, as given by the following relation for the linear alkyl chains with number of carbon atoms, n_c [15]:

$$L_{hc} = (1.5 + 1.265 \cdot n_c) \text{ Å}$$
$$\simeq R_{hc} \tag{2}$$

one can determinate whether the hydrocarbon core is spherical or not from the relation given in equation (1).

The data for various ionic and nonionic (Triton-X-100) micelles are given in table 1 as examples. It is conclusively observed that ionic micelles (e.g. n-$C_8H_{17}SO_4Na$, n-$C_{10}H_{21}SO_4Na$, n-$C_{11}H_{23}SO_4Na$ and n-$C_{12}H_{25}SO_4Na$) are all spherical in H_2O as solvent, without any added constraints on the free energy of micelle formation, ΔG_m^0 [25]:

$$\Delta G_m^0 = \Delta G_{m, \text{hph}}^0 + \Delta G_{m, \text{el}}^0 + \Delta G_{m, \text{hyd}}^0 \tag{3}$$

where subscript (hph) arises from the hydrophobic interactions (which being negative are the driving force for the aggregation process); (el) arises from the electrostatic repulsion between the ionic polar groups at the micelle-water interface (being positive it hinders the growth of micelles); (hyd) arises from the hydration factor and being positive also hinders the growth of micelles (especially in the case of nonionic micelles). If constraints such as added electrolyte concentration are increased, i.e. NaDDS in 0.6 M NaCl, then we find for this case that it cannot be a spherical micelle, because $R_{hc} > L_{hc}$. These examples also show that the 'cavity' in which micelles are present in the solvent [26] are spherical when no constraints are applied to ionic micelles. This is reasonable since a spherical cavity would give minimum free energy in the system due to symmetry, in contrast to ellipsoidal shapes (where curvature would be varying).

In the further analysis of Triton-X-100 [9] assuming a prolate ellipsoid (with aggregation number 143),

the calculated radius of a sphere of equal volume ($= (ab^2)^{1/3}$), where a is the semiaxis of the long dimension, and b is the semiaxis of the short dimension ($L_{hp} = 10$ Å $= 1.0$ nm). This gives the value of $b = 123$ Å (12.3 nm). Similarly for the case of oblate ellipsoid, one obtains the radius of a sphere of equal volume ($= (a^2b)^{1/3}$). Assuming $b = 10$ Å gives $a = 35$ Å (3.5 nm) (fig. 1).

Hydrophilic region: In the case of nonionic micelles, the hydrated ethyleneoxide adducts which would be situated at the outer surface of the hydrocarbon core would be appreciable hydrated [6, 9, 27]. This hydrophilic region would thus consist of the ethyleneoxide adducts plus the water of hydration, and would thus contribute an appreciable magnitude of volume to the total micelle volume. On the other hand, in the case of ionic micelles, like those of sodium n-dodecyl sulfate, NaDDS, this hydrophilic region would contribute comparatively little due to the low hydration and only the polar site for hydration water to be trapped as hydration. In the analysis of Triton-X-100 [9] it was assumed that the ellipsoid hydrocarbon core would be enlarged by simply adding the length of the ethyleneoxide adduct to the a and b axes of the hydrophobic core. The ethyleneoxide adduct was taken to be fully meander type which would contribute a length of 17 Å. This consideration would thus give the axes for the whole micelle as: prolate = 27 Å (2.7 nm) and 140 Å (14 nm), while for the oblate the axes were 27 Å and 52 Å (5.2 nm) [9]. The amount of water of hydration in Triton-X-100 was further calculated by different methods. The *first method* was based on the estimation of the viscosity factors of Simha [28] from the calculated axial ratios of the prolate and oblate ellipsoids. From the relation between the measured intrinsic viscosity $|\eta|$, and Simha's viscosity factor, v, one can estimate the degree of hydration:

$$|\eta| = v(v + \delta v_0) \tag{4}$$

where v is the partial specific volume of the micelle, v_0 is the specific volume of solvent, and $\delta = H_2O/$amphiphile (g/g) (hydration factor). The magnitude of v is 2.5 for spherical molecules, while it is greater than 2.5 for ellipsoid shapes [28]. The *second method* for the hydration was suggested to be through simple geometrical considerations [9]. This procedure was used to estimate the difference in the volume of ellipsoid (prolate or oblate) and the volume occupied by the amphiphile per micelle which gives a volume that was suggested to be equivalent to that occupied by the trapped water.

Table 2. Physical parameters for prolate (pr) and oblate (ob) Triton-x-100 micelles at different temperatures (this study)

Temp.	$|\eta|$ (cm³/g)	N	Semi axis Å		Ratio		Simha factor (ν)		V^m ($\times 10^{-3}$)		δ from $|\eta|$		δ from volume	
			pr a	ob b	pr a/b	ob a/b	pr	ob	pr	ob	pr	ob	pr	ob
25	4.5	150	147	53	5.4	2	6.4	2.9	448	318	−0.2	0.64	1.9	1.1
30	5.3	170	165	55	6.1	2.1	7.2	2.9	503	347	−0.2	0.92	1.9	1.0
35	6.4	209	198	60	7.3	2.2	8.8	2.9	604	401	−0.2	1.3	1.8	0.9
40	6.8	260	242	64	9	2.4	11.8	3.1	739	470	−0.33	1.3	1.8	0.8
45	7.3	322	296	70	11.6	2.6	15.7	3.2	903	551	−0.44	1.4	1.7	0.7
[Ro]	5.3	143	141	52	5.2	1.9	6.1	2.8	431	309	−0.04	1.0	1.9	1.1

[monomer mol wt = 628; mol wt of hydrocarbon (hc) part = 189; $V^l_{hc} = \dfrac{189 \cdot 10^{24}}{0.87 \cdot 6.023 \cdot 10^{23}} = 363$ Å³;

density of hydrocarbon region = 0.87 g/cm³; b (pr) = b (ob) = 27 Å]
[Ro data from Robson & Dennis (1977) where temperature is not given]

Table 3. Aggregation number, $|\eta|$ and other parameters for nonionic micelles (NP-10/13/18) (25 °C)

Surfactant	N	L_p	$|\eta|$	ν		Semiaxis		a/b		$V^m \times 10^{-3}$		δ from			
				pr	ob	pr a/b	ob a/b	pr	ob	pr	ob	$|\eta|$		volume	
												pr	ob	pr	ob
NP-10	142	18	3.8	6.4	2.8	150/28	54/28	5.4	1.9	493	347	−0.3	0.5	2	1.2
NP-13	83	23.3	4.9	3.8	2.7	100/33	51/33	3	1.5	455	356	0.4	0.9	3	2
NP-18	50	32.2	6.0	2.9	2.6	80/42	54/42	1.9	1.3	595	513	1	1.2	5.5	4.7

The present study reports the pertinent data needed for such an analysis for various micellar systems, e.g. Triton-X-100 (at different temperatures), Nonyl-phenol with ethyleneoxide-ethanol adducts (NP-10, NP-13 and NP-18, with 10, 13 and 18 numbers of ethyleneoxide units, respectively) and NaDDS (with varying concentration of added NaCl).

Experimental

Materials: Triton-X-100, Nonylphenol (NP) with varying ethyleneoxide units (NP-10, NP-13, NP-18) were used as supplied. Sodium-n-dodecyl sulfate (NaDDS) was used as supplied by B.D.H.U.K. (purity >99%).

Viscosity: For viscosity measurements a Coulter visco-meter was used. The capillary diameter was 0.1 mm. The relative viscosity, $\eta_r = t/t_0$, where t is the time for the solution and t_0 is the time for the solvent, was determined for each concentration (c) of surfactant solution. From the plots of $\dfrac{\eta_{sp}}{c} = \dfrac{\eta_r - 1}{c}$, the values of intrinsic viscosity, $|\eta|$, were obtained from the limit $= \left(\dfrac{\eta_{sp}}{c}\right)_{c \to 0}$. All plots of ($\eta_{sp}/c$) versus c were linear (within the experimental accuracy).

Results and discussion

A. Triton-X-100 micelles: The number average aggregation number, N_n, of Triton-X-100 was measured by membrane osmometry [29] which is given together with the measured intrinsic viscosity, $|\eta|$, data as a function of temperature in table 2. For the sake of comparison we also give the (average) data on the weight average aggregation number, N_w, and the intrinsic viscosity as used by [9] in their analysis. It is of interest to note that the ratio N_w/N_n is very close to 1, as has been reported by previous investigators [12], for other micellar systems.

It is seen that the prolate ellipsoid shape is unacceptable since the hydration values are negative, at all temperatures. We will argue that under these gross assumptions and approximations, the magnitude of hydration should be very critical, and must be in agreement in numerical value with the data reported by other methods of analysis before this model can be of general acceptance. The value 1.3 g H_2O/g surfactant ($\sim$5 moles H_2O/unit ethyleneoxide) is reasonable and in agreement with other results [6].

Table 4. Aggregation number (N_n), $|\eta|$ and other physical parameters for NaDDS micelles at 40 °C in varying NaCl aqueous solutions

NaCl(M)	N_n	v pr/ob	$|\eta|$ cm³/g	Semiaxis		a/b		$V^m \times 10^{-3}$		δ from $|\eta|$ gH₂O/g NaDDS		δ from value gH₂O/g NaDDS	
				pr a	ob a	pr	ob	pr	ob	pr	ob	pr	ob
0.1	103	2.76/2.56	3.64	39.7	30	1.7	1.2	90	88	+0.45	0.56	1.0	0.96
0.2	112	2.83/2.57	3.37	42.6	31.1	1.83	1.3	97	94	0.33	0.45	0.98	0.93
0.3	117	2.88/2.6	3.38	44.2	31.7	1.9	1.36	101	98	0.31	0.45	0.96	0.91
0.4	130	2.96/2.6	3.42	48.1	32.9	2.06	1.41	109	106	0.29	0.46	0.94	0.88
0.5	138	3.04/2.6	3.78	51	33.8	2.19	1.45	110	112	0.38	0.59	0.92	0.86
0.6	183	3.5/2.66	3.83	65.4	37.9	2.81	1.63	149	140	0.16	0.28	0.87	0.77
0.7	346	5.67/2.88	4.06	117.8	49.7	5.06	2.13	268	241	−0.15	0.55	0.79	0.62

[mol wt of monomer = 288; L_{hc} = 17 Å; ϱ_{hc} = 0.749 g/cm³ = ϱ for n-C₁₂H₂₆; V_{hc}^l = 350 Å⁻³; L_p = 6.5 Å
b(pr) = b(ob) = 10.7 + 6.6 = 23.3 Å; Length of ($-OSO_3Na$) = 6.6 Å (from X-ray data).

The amount of entrapped water was also estimated from volume considerations [9]. We find that either prolate or oblate ellipsoid give reasonable hydration values, $\delta = 1 - 2$ ($3 - 7$ moles H₂O/mole ethyleneoxide unit). However, it is of interest to note that the degree of hydration decreases in the case of the oblate model, which is what one would expect as the temperature increases [6, 27, 29]. These data thus indicate that by both methods the *oblate* ellipsoid is the most reasonable shape for these micelles at a range of temperature.

B. The micelle systems of nonylphenoxyethoxyethanols, e.g. NP-10, NP-13, NP-18 (where 10, 13 and 18 are the number of ethoxy units) were analyzed in the same manner as for Triton-X-100. In the calculations the following data were used (table 3):

$$L_{hc} = 11 \text{ Å} \quad \text{for NP-10, NP-13, NP-18}$$
$$(L_{hc} \text{ for Triton-X-100 was 10 Å})$$
$$L_p = \left(\frac{17}{9.5} \times (\text{number of ethyleneoxide units})\right).$$
$$(\text{The length of 9.5 units in}$$
$$\text{Triton-X-100 was 17 Å}).$$

The magnitude of N decreases as the number of ethyleneoxide units increases, which is a well accepted property of nonionic micelles [6, 27, 29]. The magnitude of $|\eta|$ is found to increase with increasing ethyleneoxide units. In the case of Triton-X-100 (table 2) we observed a similar trend with increasing temperature. The estimated δ value for NP-10 is negative for the prolate shape while it is positive for the oblate. We can thus conclude that for NP-10 the most probable shape of micelles is oblate, as was also found for Triton-X-100. On the other hand, the δ values for NP-13 and NP-18 are positive for both prolate and oblate ellipsoids, and hence one cannot easily conclude which of the two shapes is preferred. However, the δ values for prolate shape calculations are too low (0.4 g H₂O/g NP-13 ~ 1 mole H₂O/ethyleneoxide unit), which allows one to conclude that an oblate (with hydration 3 mole H₂O/ethyleneoxide unit) ellipsoidal shape is preferred. The volume method leads to very large δ values which are unacceptable (12 mole H₂O/ethyleneoxide units), and can thus be shown to be inappropriate for such considerations.

C. Sodium dodecyl sulfate (NaDDS) micelles: The shape and size analysis of ionic micelles is of great interest, since in these systems the contribution of the hydration of the polar groups would be much smaller to the hydrocarbon core than in the case for nonionic micelles, and this would lead to a more useful system for analysis of the model as described herein.

The intrinsic viscosity, $|\eta|$, data of NaDDS in systems with varying concentration of NaCl is given in table 4 at 40 °C. The number average, N_n, aggregation number under the same conditions has been published elsewhere [12]. Our $|\eta|$ data for 25 °C in 0.1 M NaCl were in agreement with the reported literature [30], i.e. $3.36 \cdot 10^{-2}$ dl/g. The $|\eta|$ data at 40 °C also compare with the literature data [18], within the experimental accuracy.

The NaDDS data were analyzed in the same manner as described above for Triton-X-100 (table 4). The data used for various parameters were:

$$L_p = \text{the length of } (-OSO_3N_a) \text{ group was}$$
$$\text{estimated to be 6.6 Å from the X-ray data.}$$

$$b(\text{pr}) = b(\text{ob}) = L_{hc} + 6.6 = 16.7 + 6.6 = 23.3 \text{ Å}.$$

L_{hc} was calculated from equation (1), using $n_c = 12$.

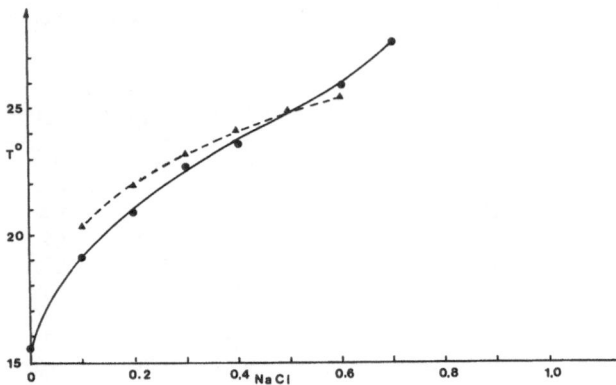

Fig. 2. Krafft point versus added NaCl (mol/dm³) concentration for NaDDS solutions: (●) NaDDS from Serva, Germany (this study, purity 99); (▲) References 7 and 14 (ca. 66% pure)

Before discussing the various parameters, it is worth analysing the variation of N_n and $|\eta|$ with increasing NaCl concentration. The abrupt change in N_n at NaCl ∼ 0.5 M was also reported by other investigators using the quasi-elastic light-scattering method [7]. The magnitude of $|\eta|$ also shows an abrupt increase when NaCl concentration is ∼0.5 M. This is expected since the addition of electrolyte lowers the Krafft point of NaDDS, figure 2. However, the electroviscous effects would have to be considered at very low electrolyte concentrations. These data show that at some temperature about 15° below Krafft point the micellar system shows a change from spherical (without constraints) to non-spherical (ellipsoidal shape) transitions. It is also clear from the simple considerations above, where data in table 1 conclusively indicate, that the transition from spherical to ellipsoidal shape is a reasonable conclusion. The data show that there is minimal micellar growth at low salt concentrations in NaDDS systems at 40 °C. The volume method gives values which are too high for δ (∼15 moles H_2O/monomer), and we can thus disregard the 'naive' volume method.

The analysis of data in table 4 indicates that the magnitude of δ is positive for both prolate and oblate ellipsoids, with the exception of 0.7 M NaCl solution, where the former ellipsoidal shape gives negative hydration values. Under these circumstances it may seem difficult to interpret the data, in contrast to the conclusive results presented for the system Triton-X-100. However, we will revert to the initial case where it was argued that spherical micelles are found to be present when no constraints are applied to the NaDDS systems (table 1). The axial ratio a/b would

increase moderately when the magnitude of N increases from 60 to 100. It is thus reasonable to conclude that an oblate ellipsoid is the preferred shape since the magnitudes of a/b increase rather smoothly. The very large axial ratios for the prolate ellipsoids are not acceptable, since the aggregation number increases very little from the spherical shape in pure water (table 1), where $a/b = 1$.

Conclusions

In summary, if one assumes that the hydrophobic core of a micelle is ellipsoidal, then the most consistent analysis of hydration is obtained if one assigns oblate shapes. This was found to be valid for Triton-X-100, at varying temperatures. However, it has recently been suggested [31] that temperature effect on nonionic micelles may also arise from other effects such as critical fluctuations of size. In these systems prolate shape considerations gave negative hydration values, which cannot be acceptable. The data of other nonionic micelles, e.g. NP-10, NP-13 and NP-18, were not as conclusive. However, the analyses of latter systems did provide an indication that useful hydration values were found for oblate ellipsoids, while prolate shapes were not acceptable.

The extensive data of NaDDS were analyzed using the same procedure. The values of hydration were positive for both prolate and oblate ellipsoids, with the exception of one system (i.e. for 0.7 M NaCl). The present data are not sufficient to allow one to determine the most preferred shape. However, a/b ratio is 5 for 0.7 M NaCl for the prolate, while it is 2.1 for the oblate ellipsoid. The value of $a/b = 5$ is rather unacceptable, since it is more reasonable to accept that the axial ratio increases to 2 from 1.2, as the aggregation number increases from 103 to 306. The latter would suggest oblate ellipsoids would be more likely from these arguments, while other workers [14] have suggested prolate ellipsoids for NaDDS micelles.

References

1. Debye P, Anacker EW (1951) J Phys Chem 55:644
2. Kushner LM, Duncand BC, Hoffman TI (1952) J Res Nat Bur Stand 49:85
3. Kushner LM, Hubbard WD, Parker RA (1957) J Res Nat Bur Stand 59:13
4. Courchene WL (1964) J Phys Chem 68:1870
5. Mysels KJ, Princen LH (1959) J Phys Chem 63:1696
6. Birdi KS (1974) Kolloid-Z u Z Polymere 252:551

7. Mazer NA, Carey MC, Benedek GB (1977) In: Mittal KL (ed) Micellization, Solubilization and Microemulsions. Plenum Press, New York
8. Leibner JE, Jacobus J (1977) J Phys Chem 81:130
9. Robson RJ, Dennis EA (1977) J Phys Chem 81:1027
10. Menger FM (1979) Acc Chem Res 12:111
11. Nicoli DF, Dawson DR, Offen HW (1979) Chem Phys Lett 66:291
12. Birdi KS, Dalsager SU, Backlund S (1980) JSC, Faraday I 76:2035
13. Briggs J, Nicoli DF, Ciccolello R (1980) Chem Phys Lett 73:149
14. Missel PJ, Mazer NA, Benedek GB, Young CY (1980) J Phys Chem 84:1044
15. Tanford C (1980) The Hydrophobic Effect, 2nd ed. Wiley, New York
16. Ikeda S, Hayashi S, Imae T (1981) J Phys Chem 85:106
17. Porte G, Appell J (1981) J Phys Chem 85:2511
18. Corti M, Degiorgio V (1981) J Phys Chem 85:711
19. Dill K, Flory PJ (1981) Proc Natl Acad Sci, USA 78:676
20. Fromherz P (1981) Chem Phys Lett 77:460
21. Briggs J, Dorshow RB, Bunton CA, Nicoli DF (1982) J Chem Phys 76:775
22a. Chattoraj DK, Birdi KS, Dalsager SU (1982) In: Mittal KL, Fendler EJ (eds) Solution Behavior of Surfactants. Plenum Press, New York
22b. Birdi KS, Stenby EH, Chattoraj DK (1983) In: Mittal KL, Lindman B (eds) Surfactants in Solution, Vol 2. Plenum Press, New York, pp 745 – 754
23. Birdi KS (1982) Finnish Chem Letts 6 – 8:142
24. Huisman JF (1964) Proc Kon Ned Akad Wetensch, Thesis, Utrecht University, Utrecht, Holland
25. Birdi KS (1976) In: Mittal KL (ed) Micellization, Solubilization and Microemulsions. Plenum Press, New York
26. Ben-Naim A (1980) The Hydrophobic Interactions. Plenum Press, New York
27. Schick MJ (1967) Nonionic Surfactants, Vol 1. Marcel Dekker, New York
28. Tanford C (1961) Physical Chemistry of Macromolecules. Wiley, New York
29. Birdi KS (1975) In: Mittal KL (ed) Colloidal Dispersions and Micellar Behavior. ACS Symp Series, Washington DC USA
30. Parker RA, Wasik SP (1958) J Phys Chem 62:967
31. Corti M, Minero C, Degiorgio V (1984) J Phys Chem 88:309

Received July 2, 1984;
accepted October 15, 1984

Author's address:

K. S. Birdi
Fysisk-Kemisk Institut
The Technical University of Denmark
Building 206
DK-2800 Lyngby (Denmark)

Some observations on liquid crystallinity in barium surfactant systems

A. Khan[a], K. Fontell[a], and B. Lindman[b]

Division of Physical Chemistry 2[a] & 1[b], Chemical Centre, University of Lund, Lund (Sweden)

Abstract: The phase diagrams of barium surfactants in two and three component systems are investigated by water deuteron (^{2}H) NMR and polarizing microscopic studies. Water-free barium di-2-ethylhexylsulphosuccinate (BaA_2) has two-dimensional hexagonal structure and forms a reverse hexagonal liquid crystalline phase (F) with water at low water contents ($\lesssim 16$ wt%). At the high water content limit the F phase is in equilibrium with a very dilute aqueous solution (L_1). The solubility of BaA_2 does not increase significantly with increasing temperature. The two-phase region, $L_1 + F$, transforms, on heating, at 316 K (three-phase line) to two immiscible liquid phases, $L_1 + L_2$, which were found not to mix up to the range where decomposition starts.

Barium octylsulphate (BaOS) is virtually insoluble in water and does not form any liquid crystalline phase with water between 298 and 323 K. Addition of decanol to the aqueous system leads to the formation of the lamellar liquid crystalline phase, yet incompletely characterized. However, it appears that the lamellar phase exists at low water contents and probably occupies a small area in the phase diagram. The phase diagrams of the barium surfactant systems are compared with those of the corresponding systems with Mg^{2+} and Ca^{2+} counterions and the results are discussed in the light of electrostatic theory.

Key words: Barium di-2-ethylhexylsulphosuccinate, barium octylsulphate, liquid crystals, phase diagram, ^{2}H NMR.

Introduction

Recently, we have shown that the stability of the lamellar liquid crystalline phase is reduced considerably by replacing a monovalent counterion with a divalent one in anionic surfactant systems [1 – 4]. The dependence of the swelling capability of the lamellar liquid crystalline phase on the valency of the counter ion appears to be a general characteristic for the ionic surfactant systems.

This finding may be understood qualitatively from the electrostatic theory [5 – 6] and Monte Carlo simulations [7]. The swelling capability of the lamellar liquid crystalline phase has been shown to be due to repulsion between charged surfactant layers [5]. The fraction of divalent counterions bound to the charged surfaces is higher than that of the monovalent ions, leading to a smaller interlamellar repulsion, thus causing a reduction of the swelling.

Different divalent counterions are again found to effect the phase equilibria of the system somewhat differently, e.g. phase extension, thermal stability of the liquid crystalline phases, and formation and/or disappearance of phases.

The studies reported so far on the surfactant systems with divalent counterions are very few and limited to calcium and magnesium. Here we present the binary phase diagram (composition vs temperature) of the system barium di-2-ethylhexylsulphosuccinate (BaA_2) – ^2H_2O. We will also make a preliminary report on the isothermal ternary system of barium octylsulphate (BaOS)-decanol-^2H_2O.

Materials

Barium di-2-ethylhexylsulphosuccinate (BaA_2) was prepared by (a) precipitation and (b) ion exchange methods from sodium di-2-ethylhexylsulphosuccinate (NaA) which was purified from absolute methanol prior to its use.

a) BaA_2 was precipitated by adding $BaCl_2$ (20% in excess of stoichiometric amount) to NaA in methanolic solution (CH_3OH/H_2O : 1/4 (v)). The precipitate was separated and purified by repetitive washing of the solid with CH_3OH/H_2O mixture followed by centrifugation. (Separation of the precipitate by filtration is difficult). Finally, the solid was dried under vacuum at 333 K.

b) About 0.1 M NaA in methanolic solution $(CH_3OH/H_2O:1/1(V))$ was eluated through a column containing the ion-exchange resin Amberlite IR-120 in its hydrogen-ion form. The eluate was neutralized with $Ba(OH)_2$ to pH ~8. The filtered, clear solution was lyophilized.

BaA_2 prepared by (a) and (b) was recrystallized from absolute methanol and the crystals were dried under vacuum at 333 K.

BaA_2 obtained by (a) or (b) produced identical results in the study of the phase diagram.

c) Barium octylsulphate (BaOS) was precipitated by adding $BaCl_2$ (20% in excess of stoichiometric amount) to an aqueous solution of sodium octylsulphate. The precipitate was separated by filtration and purified by repetitive washings. Finally, the solid was dried under vacuum at 30°C. BaOS was recrystallized from absolute methanol followed by drying the crystals under vacuum.

A concentrated solution of BaA_2 or BaOS in methanol did not show the presence of Na^+ when ^{23}Na NMR was run with a few thousand pulses whereas a sharp peak on the ^{23}Na NMR spectrum was observed with a few pulses for a corresponding solution containing NaA or sodium octylsulphate.

Sample preparations

The samples were prepared by weighing the appropriate amounts in glass tubes which were sealed-off immediately. The samples were mixed and equilibrated as described previously [3].

Methods

The phase equilibria of the surfactant systems have been studied by observing the water deuteron (^2H) NMR spectra at a resonance frequency of 15.351 MHz with a modified Varian XL-100-15 pulsed spectrometer. The experimental details and the analysis of NMR spectra are as reported previously [1].

While isotropic systems give rise to a sharp singlet in the ^2H NMR spectra, quadrupole splittings are obtained in the anisotropic liquid crystalline phases. The magnitude of the splitting (which depends on the structure of the liquid crystalline phase) is, in general, much larger in the lamellar than in the hexagonal liquid crystalline phase [8]. For a heterogeneous system consisting of two or more phases one obtains a superposition of the ^2H NMR spectra characterizing the different phases, provided that the water deuteron exchange between the phases is slow.

The texture of the liquid crystalline sample was studied at room temperature in a polarizing microscope equipped with a Koffler hot stage and its structure was inferred from the texture [9]. The change in the texture was then followed when the temperature was raised at a rate of about 2 K per minute.

X-ray low- and wide-angle diffraction studies of the liquid crystalline phases were performed in a manner previously reported [10]. The low-angle camera was a point-collimated

Kiessig camera and the X-ray radiation was nickel-filtered copper K_α. Due to the presence of barium in the samples the diffractograms were of poor quality but they could nevertheless confirm the phase assignations obtained by polarizing microscopy.

The phase diagram was constructed by ^2H NMR and polarizing microscopic studies in combination. The results are accurate to $2-3$ K on the temperature scale and 1 wt% on the concentration scale.

Results

Barium di-2-ethylhexylsulphosuccinate

The phase diagram of the system $BaA_2-{}^2H_2O$ shown in figure 1 is rather simple. It forms a reverse hexagonal phase, F, at low water content ($\lesssim16$ wt%) as the only liquid crystalline phase and water-free BaA_2 shows, like AOT (Aerosol OT) [11] hexagonal texture in the polarizing microscope. X-ray diffraction studies also show that BaA_2 has a two-dimensional hexagonal structure at room temperature. The structure is retained on adding water. The neat BaA_2 has no melting point and at about 478 K, it decomposes and turns yellow.

The liquid crystalline samples in the F phase produce ^2H NMR quadrupolar splittings and in the two-phase region, L_1+F, a quadrupolar splitting together with a sharp central peak is obtained in the ^2H NMR spectrum. A few quadrupolar splitting values (Δ) are shown in table 1. At the same surfactant $-$ water molar ratios, Δ-values obtained in

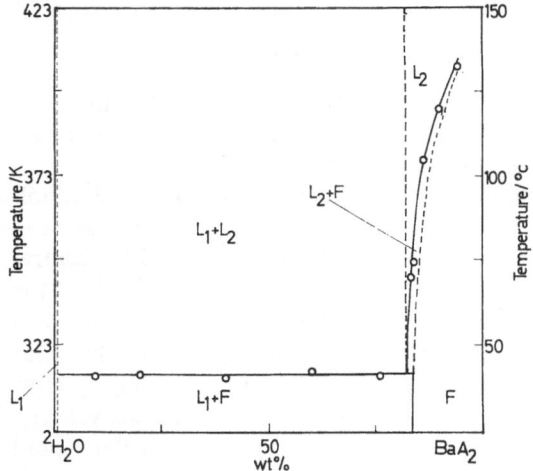

Fig. 1. Phase diagram of the barium di-2-ethylhexylsulpho-succinate-^2H_2O system
Phase notations: F, reverse hexagonal liquid crystalline phase, L_1 and L_2, isotropic solution phases

Table 1. The water deuteron (^{2}H) NMR quadrupolar splittings (Δ) obtained in samples of the liquid crystalline phase for barium di-2-ethylhexylsulphosuccinate-^{2}H$_2$O system at 300 K

Surfactant conc. (wt %)	Phase designation	$\dfrac{\Delta}{\text{kHz}}$
80.2	$F + L_1$	$2 \cdot 4$ + Isotropic peak
83.8	F	$2 \cdot 5$
86.5	F	$2 \cdot 7$
90.2	F	$3 \cdot 8$
94.5	F	$5 \cdot 9$

this system are closely the same as those obtained in the F phase of the calcium and magnesium systems [3].

BaA$_2$ is almost insoluble ($<10^{-3}$ molal) in water at room temperature and the solubility does not increase appreciably on increasing temperature. Samples in the large two phase region, $L_1 + F$, are converted, on heating, to two immiscible isotropic liquid phases, $L_1 + L_2$, at 316 K (three-phase line). The two-phase region, $L_1 + L_2$, is thermally stable up to 423 K (heated for 4 h) and there is no mixing of these two phases within this temperature range. A few samples were heated to 523 K. Above 473 K, the two phases appear to mix, but there is a substantial decomposition of the samples making a characterization of the phase equilibrium difficult.

Barium octylsulphate

Barium octylsulfate (BaOS) is insoluble in water and it does not form any liquid crystalline phase with water between 298 and 323 K.

The samples in the ternary system BaOS-decanol-^{2}H$_2$O show lamellar (mosaic) microscopic texture and produce a quadrupole splitting together with a central peak on the ^{2}H NMR spectrum. The magnitude of the quadrupole splitting ($2 - 3$ kHz) is comparable with that obtained in the water-poor region of the lamellar liquid crystalline phases of the corresponding systems with Ca^{2+} or Mg^{2+}. However, we have not yet succeeded in obtaining a homogeneous lamellar liquid crystalline sample of this system. It appears though that the lamellar liquid crystalline phase forms at low water contents and occupies a small area in the phase diagram.

Discussion

In the ternary alkaline earth octylsulphate-decanol-water system, the extension of the lamellar liquid crystalline phase towards the water corner increases in the order Mg^{2+} > Ca^{2+} > Ba^{2+} (28 moles ^{2}H$_2$O per mole surfactant ion for magnesium against 18 moles for the calcium system and not quantified for the barium system). On the other hand, the minimum amount of water necessary to form the lamellar liquid crystalline phase in the calcium and magnesium systems is approximately the same, ~ 8 moles ^{2}H$_2$O per mole surfactant ion [4].

The stability of the lamellar liquid crystalline phase in the binary systems of MgA$_2$ and CaA$_2$ with water follows the same trend as in the ternary systems while no lamellar liquid crystalline phase is found to exist in the barium system. Moreover, the cubic liquid crystalline phase that forms with Mg^{2+} and Ca^{2+} counterions is absent with Ba^{2+} ions. The only liquid crystalline phase common to these three surfactant systems is the reverse hexagonal phase. The maximum number of ^{2}H$_2$O molecules per surfactant ion (n) that the homogeneous hexagonal liquid crystalline phase can incorporate in the three systems are, respectively, $n_{\text{Mg}} = 3 \cdot 2$, $n_{\text{Ca}} = 3 \cdot 3$ and $n_{\text{Ba}} = 4 \cdot 8$. The increased capability of the lamellar liquid crystalline phase to swell and incorporate water with divalent counterions (Mg^{2+} > Ca^{2+} > Ba^{2+}) follows the order of the size of the hydrated ions [12]. Counterion binding is expected to be facilitated by a reduced distance of closest approach between counterion and headgroup; this results from a reduced hydration. It is, therefore, expected from the electrostatic approach that the interlamellar repulsion reduces (hence reduced swelling) in the sequence Mg$^{2+} \rightarrow$ Ca$^{2+} \rightarrow$ Ba^{2+}.

For the reversed hexagonal phase geometrical constraints are different. A low hydration would increase the number of water molecules that can form the hydrophilic cylinders. Furthermore, this phase is characterized by a much closer packing of polar headgroups than in the lamellar phase; hydration would be an opposing factor to this packing. Here we only note that this oversimplified reasoning is consistent with the observed trend.

Two-phase mixtures of water with MgA$_2$ ($L_1 + D$), CaA$_2$ ($L_1 + D$) or BaA$_2$ ($L_1 + F$) form, on heating, a heterogeneous system consisting of two immiscible liquids, $L_1 + L_2$. The transition temperature (three phase tie-line) increases in the order Ba < Ca < Mg. The two liquids, $L_1 + L_2$, appear not to mix within the thermal stability range of the surfactant (423 K). The coexistence of two liquid phases, $L_1 + L_2$, in binary alkali ion surfactant systems is, on the other hand, uncommon.

References

1. Khan A, Fontell K, Lindblom G, Lindman B (1982) J Phys Chem 86:4266
2. Khan A, Fontell K, Lindman B (1984) In: Mittal KL, Lindman B (eds) Surfactants in Solution, Vol 1. Plenum Press, New York, p 193
3. Khan A, Fontell K, Lindman B (1984) J Colloid Interf Sci 101:193
4. Khan A, Fontell K, Lindman B (1984) Colloids and Surfaces 11:401
5. Jönsson B, Wennerström H (1981) J Colloid Interf Sci 80:482
6. Jönsson B, Gunnarsson G, Wennerström H (1982) In: Mittal KL, Fendler EJ (eds) Solution Behaviour of Surfactants, Vol 1. Plenum Press, New York, p 317
7. Wennerström H, Jönsson B, Linse P (1982) J Chem Phys 76:4665
8. Wennerström H, Lindblom G, Lindman B (1974) Chem Scr 6:97
9. Rosevear FB (1968) J Soc Cosmet Chem 19:581
10. Fontell K (1973) J Colloid Interf Sci 43:156
11. Rogers J, Winsor PA (1967) Nature 216:477
12. Conway BE (1981) In: Ionic Hydration in Chemistry and Biophysics. Elsevier Scientific Publishing Company, Amsterdam New York, p 59

Received October 15, 1984;
accepted October 30, 1984

Authors' address:

A. Khan
Division of Physical Chemistry 2
Chemical Center
Box 124
S-22100 Lund (Sweden)

Shape transitions in aqueous micellar systems as a function of pressure and temperature

E. Ljosland[1], A. M. Blokhus[1], K. Veggeland[1], S. Backlund[2], and H. Høiland[1]

[1]Department of Chemistry, University of Bergen, N-5014 Bergen-U. (Norway)
[2]Department of Physical Chemistry, Åbo Akademi, SF-20500 Åbo (Finland)

Abstract: Viscosity, conductivity, and ultrasound velocity measurements have been used to study shape transitions of micelles. Sodium dodecylsulfate, dodecylammonium chloride, and dodecyltrimethyl-ammonium bromide all exhibit such transitions for 0.3 m aqueous solutions in the presence of 1-alcohols. The transition point is dependent upon the temperature and pressure, and on the alcohol chain length. The alcohols from hexanol upwards promote transitions to larger aggregates, the higher alcohols being the most efficient. At higher temperatures more alcohol must be added to induce shape transition. The same applies as the pressure is increased, but the effect of pressure is less pronounced.

Key words: Micelles, shape transitions, activation energy, 1-alcohols

Introduction

In aqueous solution micelles are generally thought to be spherical as long as the surfactant concentration remains close to the critical micelle concentration. Rod-like micelles may form at higher surfactant concentrations [1, 2]. Addition of a third component such as neutral salt or non-electrolytes may favour longer micellar structures, for instance rod-like micelles [3 – 6]. An increase in temperature, on the other hand, seems to favour spherical micelles [7, 8]. The effect of pressure on the shape transition point is not known, though it appears that the aggregation number of micelles decrease with pressure at least up to about 160 MPa [9 – 12].

The shape of the micelles, spheres, rod-like or disc-shaped, must be ruled by a balance between the repulsive electrostatic forces of the head group and the attractive hydrophobic forces that cause the aggregation. It thus seems reasonable that the transition point depends on the head group of the surfactant including the counter ion as well as on the length of the hydrocarbon chain. In order to investigate the effect of the head group three different surfactants with the same hydrocarbon chain length was chosen for this study, i.e. sodium dodecylsulfate, dodecylammonium chloride, and dodecyltrimethylammonium bromide. The surfactants have previously been studied by Nagara-

jan et al. [13]. From viscosity measurements they found that addition of neutral salt did promote rod-like micelles for sodium dodecylsulfate and dodecylammonium chloride. For dodecyltrimethylammonium bromide no transition to rod-like micelles was observed even at high surfactant concentrations.

In this work we have focused on the effects of adding 1-alcohols to the surfactants mentioned above. The effects of temperature and pressure on the shape transition have also been studied.

Experimental

Sodium dodecylsulfate (NaDDS) was supplied by BDH, "specially pure" grade. Dodecylammonium chloride (DAC) and dodecyltrimethylammonium bromide (DTAB) were supplied by Eastman Kodak. The alcohols 1-hexanol (C_6OH), 1-heptanol (C_7OH), and 1-octanol (C_8OH) were supplied by Fluka at their best quality. All compounds were used without further purification. Water was distilled twice immediately before use. All solutions were made by weight.

The densities of the solutions were measured by a Paar densitometer, and the viscosities by Ostwald viscometers. The ultrasound measurements were carried out by the "sing-around" technique as previously described [14]. The conductivities were measured as previously described [15]. The temperature was controlled to better than ± 0.05 K as measured by a Hewlett-Packard quartz thermometer.

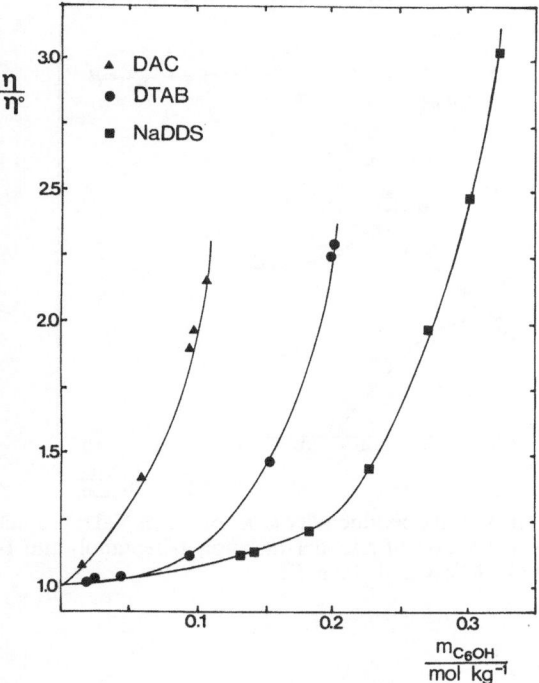

Fig. 1. The relative viscosity η/η^0 of NaDDS, DAC, and DTAB solutions as functions of added hexanol at 298.2 K and 0.1 MPa. The surfactant molality was 0.3 m in all cases and η^0 is the viscosity of the pure surfactant solutions

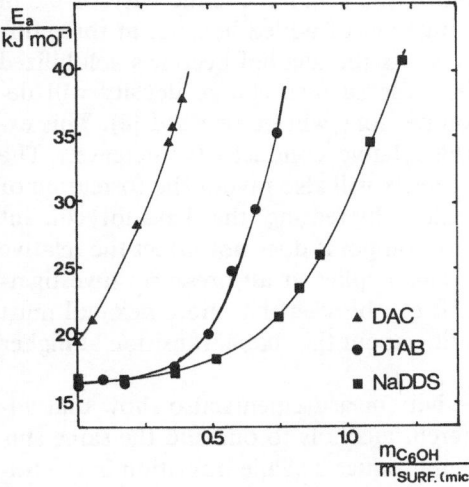

Fig. 2. The activation energy of viscous flow, E_a, as a function of added hexanol at 0.1 MPa. The micellar surfactant content has been calculated as $m_{surf(mic)} = m_{surf}^{tot} - m_c$, the total surfactant molality minus the molality at the critical micelle concentration

Results and discussion

The viscosity measurements were carried out on NaDDS, DAC, and DTAB aqueous solutions as a function of added hexanol and in the temperature region 298.2 – 318.2 K. Figure 1 shows the relative viscosities as a function of the hexanol molality. The initial surfactant molality was 0.3 m in all cases, and the measurements extend to the point where the solutions became cloudy. The dramatic increase in the relative viscosities at a certain alcohol molality indicates a shape transition [1]. While there is little doubt that the original micelles are spherically shaped, the shape of the aggregates found at higher surfactant concentrations are subject to some discussion. Data on these aggregates have been interpreted as rod-like micelles [1], oblate or prolate micelles [16], or swollen anisotropic micelles [17]. We cannot clarify this point, but from our measurements it can be seen that the amount of alcohol needed for shape transition to occur increases in the order NaDDS > DTAB > DAC. Since the hydrocarbon chain length of the surfactants is the same, this must reflect differences in the surfactant head group and its counter ion. Recent light scattering studies [18] have shown that the formation of larger aggregates, rod-like micelles, in dodecylsulfate solutions depends on the counter ion. The transition to these rod-like micelles occurs in the order Cs > K > Na > Li, i.e. the less hydrated ions are the most efficient. Ozeki [19] has used the same method to study the effect of the polar head group. The transition to rod-like micelles occurs in the order dodecylammonium > dodecyldimethylammonium > dodecyltrimethylammonium with chloride as the counter ion in all cases. It thus seems that a less hydrated counter ion and a small head group favour the such shape transtions.

The temperature dependence of the viscosity is usually expressed by an Arrhenius-type of equation:

$$\eta = A e^{E_a/RT}. \tag{1}$$

E_a is the activation energy of viscous flow. By plotting $\ln \eta$ versus $1/T$ one can determine E_a. Straight lines were observed in all cases for our systems, and the resulting activation energies have been plotted versus the hexanol molality in figure 2. At low hexanol molalities E_a is approximately 17 kJ mol^{-1} for NaDDS and DTAB solutions. This value is characteristic for water, for aqueous solutions and for spherocolloids [1, 20]. The data show that the NaDDS and DTAB micelles are spherical at low hexanol content in the temperature range 298.2 – 318.2 K. As more hexanol

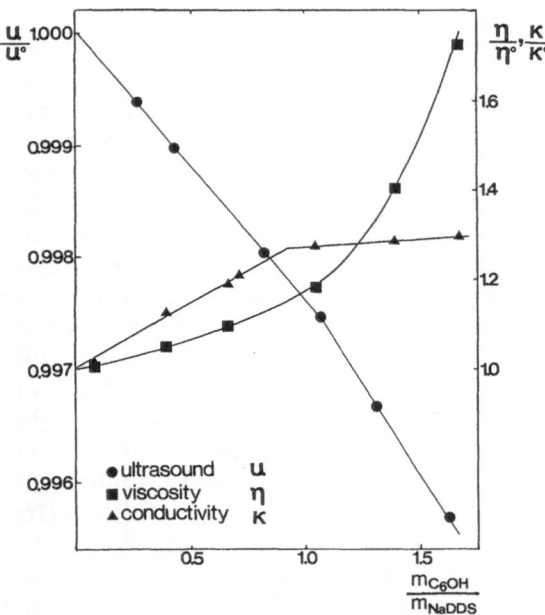

Fig. 3. The relative speed of sound, viscosity and conductivity as a function of added hexanol at 303.2 K and 0.1 MPa. The NaDDS molality was 0.2 m. (The temperature for the conductivity measurements was 298.2 K)

is added E_a increases sharply, suggesting that a shape transition takes place. The E_a value of 0.3 m DAC is slightly higher than 17 kJ mol^{-1}. However, it is presently not known whether it is due to impurities or due to the fact that we are working close to the Krafft point at the lowest temperature. In any case the E_a data also show that the shape transition occurs in the order DAC > DTAB > NaDDS.

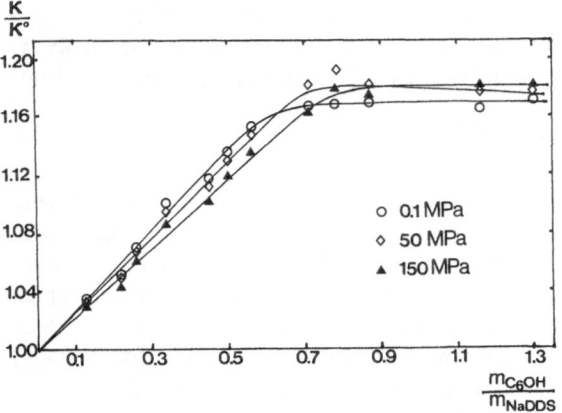

Fig. 4. The relative conductivity κ/κ^0 of 0.3 m NaDDS solutions as a function of added hexanol at three different pressures and 298.2 K. κ^0 is the conductivity of the NaDDS solution without added alcohol at the actual pressure

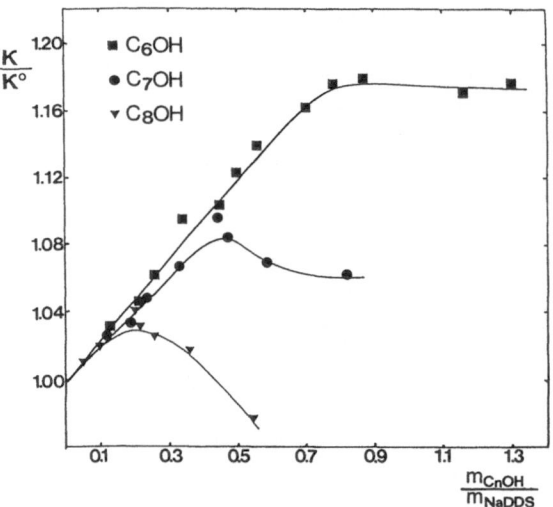

Fig. 5. The relative conductivity κ/κ^0 of 0.3 m NaDDS solutions as a function of added 1-hexanol, 1-heptanol, and 1-octanol at 298.2 K and 100 MPa

Structural changes in aqueous micellar solutions can also be seen from conductivity and ultrasound speed measurements as shown in figure 3. The relative viscosity, the conductivity, and the speed of sound have all been plotted in the same figure. The speed of sound and the conductivity curves exhibit a break at the same hexanol molality as the viscosity starts increasing. It is thus possible to use any of these methods to investigate shape transitions.

Figure 4 shows the relative conductivity of 0.3 m NaDDS as a function of added hexanol at three different pressures. As the alcohol becomes solubilized by the micelles, the surface charge density will decrease and counter ions will be released [8]. This explains why the relative conductivity increases. The lower charge density will also favour the formation of larger aggregates. Increasing the hexanol content above the transition point does not affect the relative conductivity. This applies at all pressures investigated. However, it can be seen that more hexanol must be added to bring about the shape transition at higher pressures.

The conductivity measurements also show that addition of different alcohols to one and the same surfactant solution produce a shape transition in the following order $C_8OH > C_7OH > C_6OH$; i.e. octanol is the most efficient. Figure 5 shows a typical plot of the relative conductivity of 0.3 m NaDDS solutions at 100 MPa as a function of the added amount of alcohol. The hydrocarbon chain of the solubilized alcohol will

Table 1. Alcohol content at which the sphere to rod transition takes place for 0.3 m NaDDS solutions at 298.2 K and different pressures

P MPa	$m_{hexanol}$ mol kg^{-1}-water	$m_{heptanol}$ mol kg^{-1}-water	$m_{octanol}$ mol kg^{-1}-water
0.1	0.188	0.113	0.056
50	0.218	0.128	0.060
100	0.225	0.135	0.068
150	0.233	0.162	0.078

increase the total hydrophobicity of the micelle, thus favouring large micelles. The longer the hydrocarbon chain of the alcohol, the less alcohol will be needed for the transition to take place. The alcohol content needed to induce shape transition in 0.3 m NaDDS solutions at various pressures has been presented in table 1. The effect of pressure is moderate.

The conductivity measurements also show that shape transition occurs at a lower alcohol concentration in aqueous DTAB solutions than in aqueous NaDDS as shown in figure 6 at two different pressures. It can also be seen that the effect of pressure on

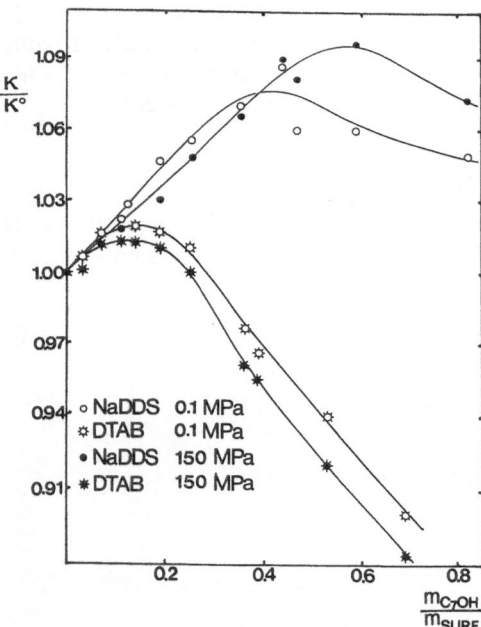

Fig. 6. The relative conductivity κ/κ^0 of 0.3 m NaDDS and 0.3 m DTAB solutions as functions of added heptanol at 298.2 K and two different pressures

the transition point of DTAB is even less than for NaDDS.

Acknowledgement

A. M. B. thanks the Norwegian Council for Science and the Humanities; and S. B. the Research Institute at Åbo Akademi for financial support.

References

1. Ekwall P, Mandell L, Solyom P (1971) J Colloid Interface Sci 35:519
2. Fendler JH (1980) J Phys Chem 84:1485
3. Mazer NA, Benedek GB, Carey MC (1976) J Phys Chem 80:1075
4. Ikeda S, Ozeki S, Hayashi SH (1980) Biophys Chem 11:417
5. Birdi KS, Dalsager SO, Backlund S (1980) J Chem Soc Faraday Trans I, 76:2035
6. Tominaga T, Stem TB, Evans DF (1980) Bull Chem soc Jpn 53:795
7. Backlund S, Høiland H, Kvammen OJ, Ljosland E (1982) Acta Chem Scand A39:698
8. Høiland H, Kvammen OJ, Backlund S, Rundt K (1984) In: Mittal KL, Lindman B (eds) Surfactants in Solution, Vol 2. Plenum Press, New York, p 949
9. Kaneshina S, Tanaka M, Tominaga T, Matuura R (1974) J Colloid Interface Sci 48:450
10. Nicoli DF, Lawson DR, Offen HW (1979) Chem Phys Lett 66:291
11. Nishikido N, Shinozaki M, Sugihara G, Tanaka M, Kaneshina S (1980) J Colloid Interface Sci 74:474
12. Nishikido N, Shinozaki M, Sugihara G, Tanaka M (1981) J Colloid Interface Sci 82:352
13. Nagarajan R, Shah KM, Hammond S (1982) Colloid Surfaces 4:147
14. Høiland H, Vikingstad E (1976) J Chem Soc Faraday Trans I, 72:1441
15. Høiland H (1974) J Chem Soc Faraday Trans I, 70:1180
16. Zana R, Picot C, Duplessix R (1983) J Colloid Interface Sci 93:43
17. Candau S, Zana R (1981) J Colloid Interface Sci 84:206
18. Missel PJ, Mazer NA, Carey MC, Benedek GB (1984) In: Mittal KL, Fendler EJ (eds) Solution Behavior of Surfactants. Theoretical and Applied Aspects. Plenum Press, New York, p 1
19. Ozeki S (1982) Hyomen 20:632
20. Ekwall P, Holmberg P (1965) Acta Chem Scand 19:573

Received June 26, 1984;
accepted October 20, 1984

Authors' address:

E. Ljosland
Department of Chemistry
University of Bergen
N-5000 Bergen (Norway)

The structure of adsorbed monolayers. The surface chemical bond

G. A. Somorjai and B. E. Bent

Materials and Molecular Research Division, Lawrence Berkeley Laboratory and Department of Chemistry, University of California, Berkeley, California (U.S.A.)

Abstract: Results from new electron and ion spectroscopies with surface sensitivities of less than 1% of a monolayer have greatly increased our knowledge of structure and bonding at the solid-vacuum interface. We present here a case-history of some of the major conclusions of these surface structural studies. Results on metal single crystal surfaces of surface composition using Auger electron spectroscopy, bond energetics using thermal desorption spectroscopy, and bonding geometries using low-energy electron diffraction and high-resolution electron energy loss spectroscopy are highlighted. A wide variety of bonding geometries and bond energies are found – one of the major reasons for the diverse reactivity and chemistry of surfaces. Recent work combining vacuum, surface science techniques with studies at solid-gas and solid-liquid interfaces shows promise in extending our understanding of surface structure and bonding to these interfaces.

Key words: Surface bond, surface structure, monolayers, single crystals

1. Introduction

During the last fifteen years new techniques developed by modern surface science have permitted a molecular level scrutiny of the surface monolayer [1]. Table 1 lists many of the techniques utilized most frequently in molecular surface science studies. Most of these surface science techniques utilize electron or ion scattering and require a change of density at the interface for surface sensitivity. As a result, most studies have been performed at the solid-vacuum and solid-gas interfaces and not at solid-liquid interfaces.

Using these techniques, the atomic surface structures of clean surfaces and adsorbate monolayers have been determined. Surface composition can be verified to better than 1% of a monolayer (10^{13} atoms/cm^2 or less). The oxidation states of surface atoms can also now be verified.

This paper attempts to provide a summary of what has been learned about the structure of adsorbed monolayers and about the surface chemical bond from molecular surface science. While the surface chemical bond is less well understood than bonding of molecules in the gas phase or in the solid state, our knowledge of its properties is rapidly accumulating. The information obtained also has great impact on many surface science based technologies, including heterogeneous catalysis and electronic devices. It is hoped that much of the information obtained from studies at solid-gas interfaces can be correlated with molecular behavior at solid-liquid interfaces.

1.1 Surface science technology

Figure 1 shows the experimental geometry that is usually used in modern surface science studies [1]. A

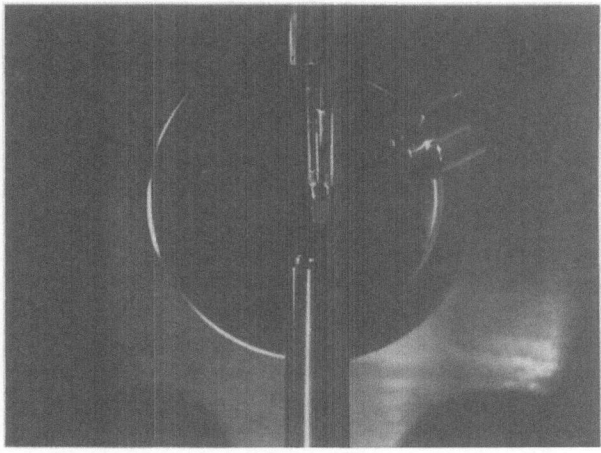

Fig. 1. Small surface area sample mounted in an ultra-high vacuum chamber prepared for surface studies

Table 1. Surface characterization techniques used to determine the structure and composition of solid surfaces. Adsorbed species present at concentrations of 1% of a monolayer can be readily detected

Surface analysis method	Acronym	Physical basis	Type of information obtained
Low-energy electron diffraction	LEED	Elastic backscattering low-energy electrons	Atomic surface structure of surfaces and of adsorbed gases
Auger electron spectroscopy	AES	Electron emission from surface atoms excited by electron, X-ray, or ion bombardment	Surface composition
High-resolution electron energy loss spectroscopy	HREELS	Vibrational excitation of surface atoms by inelastic reflection of low energy electrons	Structure and bonding of surface atoms and adsorbed species
Infrared reflection spectroscopy	IRS	Vibrational excitation of surface atoms by adsorption of infrared radiation	Structure and bonding of adsorbed gases
X-ray and ultraviolet photoelectron spectroscopy	XPS UPS	Electron emission from atoms	Electronic structure and oxidation state of surface atoms and adsorbed species
Ion scattering spectroscopy	ISS	Inelastic reflection of inert gas ions	Atomic structure and composition of solid surfaces
Secondary ion mass spectroscopy	SIMS	Ion beam induced ejection of surface atoms as positive & negative ions	Surface composition
Extended X-ray absorption fine structure analysis	EXAFS	Interference effects during X-ray emission	Atomic structure of surfaces and of adsorbed species
Thermal desorption spectroscopy	TDS	Thermally induced desorption or decomposition of adsorbed species	Adsorption energetics/ composition of adsorbed species
Solid state nuclear magnetic resonance	Solid-state NMR	Nuclear magnetic resonance on samples with areas of 1 m^2 or larger	Atomic and molecular composition, structure

sample, usually a single crystal of about 1 cm^2 surface area, is enclosed by an ultra-high vacuum chamber, and there are various surface analysis techniques surrounding it. Surface cleaning is carried out by ion bombardment and the surface orientation and structure are tested by low energy electron diffraction. Electron spectroscopies, including photoelectron spectroscopy and Auger electron spectroscopy, determine the surface composition and the oxidation state of surface atoms. High-resolution electron energy loss spectroscopy is used to study the vibrational structure of atoms or molecules adsorbed on surfaces.

Figure 2 shows the power of a single electron beam in determining many of the surface chemical properties [2]. An incident electron beam of a few hundred or a thousand electron volts in energy yields an elastically scattered fraction that is used for low-energy electron diffraction and surface crystallography to determine the precise location of atoms — their bond distances and bond angles [3]. Very minute energy losses in the milli-electron volt range result from ex-

citation of surface vibrations and provide an assignable surface vibrational spectrum [4]. Energy losses due to inner shell excitation or deexcitation processes give rise to photoelectron spectra [5] and Auger electron spectra [6].

1.2 Surface composition

Knowing the surface composition is necessary for surface structure determinations. An example of the Auger electron spectra used to determine surface composition is exhibited in figure 3 for a silver-gold alloy [7]. From the Auger peak height ratios, the silver to gold surface atom ratios were determined. It is found that there is surface segregation of one of the constituents, in this case silver, due to surface thermodynamic reasons. The surface composition is adjusted to minimize the total surface free energy, thereby segregating to the surface that constituent of lower surface energy, in this case silver. Surface composi-

ENERGY DISTRIBUTION OF SCATTERED ELECTRONS FROM
A c(4×2) MONOLAYER OF C$_2$H$_3$ ON Rh(III) AT 300K

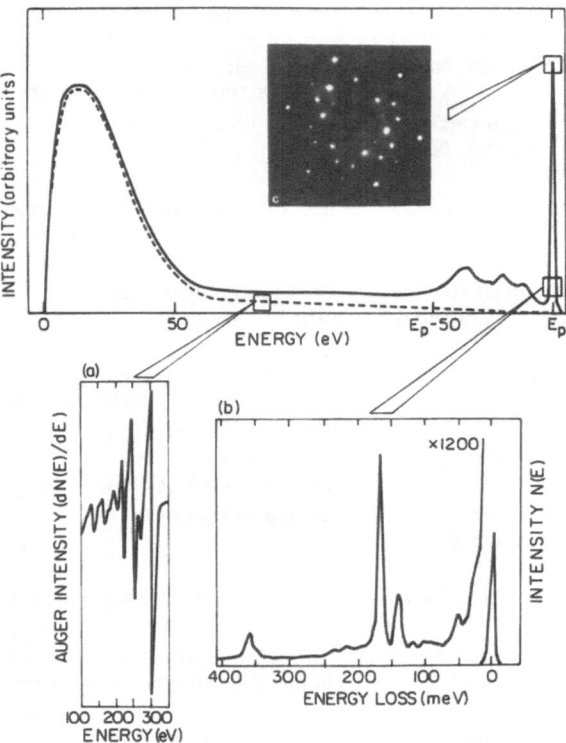

Fig. 2. Experimental number of scattered electrons, $N(E)$, of energy, E, versus electron energy for a Rh(111) surface covered with a monolayer of ethylidyne species (CCH$_3$) — the stable, room temperature structure of chemisorbed ethylene. Boxes and inset figures show how particular scattered electrons are used in (a) Auger electron spectroscopy, (b) high resolution electron energy loss spectroscopy and (c) low energy electron diffraction

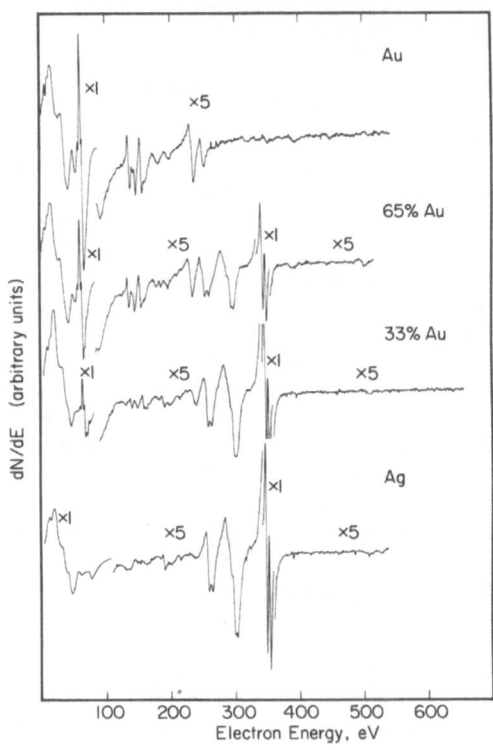

Fig. 3. Typical Auger spectra from pure Au, two alloys, and pure Ag

tion studies as a function of depth (depth profile analysis) clearly reveal changes of surface composition layer by layer from the surface inward. This is shown in figure 4. The surface layer has excess of silver relative to the bulk silver-gold alloy composition, the second layer has excess gold, the third layer has excess silver again, and by the fourth layer the bulk composition is achieved.

Figure 5 shows the regular solution and ideal solid solution equations that govern surface segregation for two component systems [8]. Here σ_1 and σ_2 are the surface tensions of the pure components 1 and 2, X_1^s and X_2^s are the atom fractions of the two components at the surface, and X_1^b and X_2^b are the atom fractions of the two components in the bulk. Ω is a regular solution parameter that is proportional to the heat of

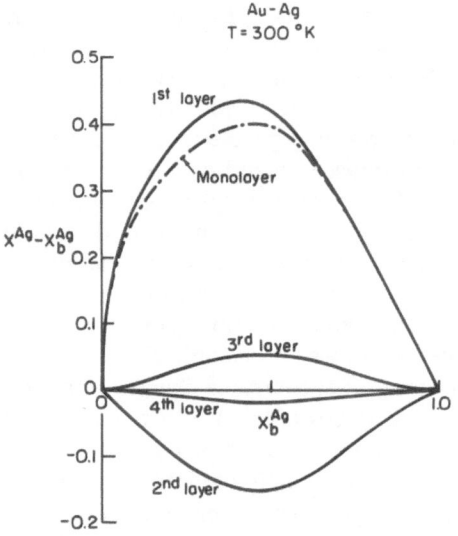

Fig. 4. Surface excess of silver as a function of bulk composition in silver-gold alloys

For an ideal solid solution:

$$\frac{x_2^s}{x_1^s} = \frac{x_2^b}{x_1^b} \exp\left[\frac{(\sigma_1 - \sigma_2)a}{RT}\right]$$

For a regular solid solution:

$$\frac{x_2^s}{x_1^s} = \frac{x_2^b}{x_1^b} \exp\left[\frac{(\sigma_1 - \sigma_2)a}{RT}\right] \exp\left\{\frac{\Omega(l + m)}{RT}\left[(x_1^b)^2 - (x_2^b)^2\right]\right.$$

$$\left.\frac{\Omega l}{RT}\left[(x_2^s)^2 - (x_1^s)^2\right]\right\}$$

where Ω = regular solution parameter = $\dfrac{\Delta H_{mixing}}{x_1^b \cdot x_2^b}$

l = fraction of nearest neighbors in surface layer.

m = fraction of nearest neighbors in adjacent layer.

Fig. 5. The ideal and regular solid solution models that predict surface segregations of the constituents with lower surface free energy

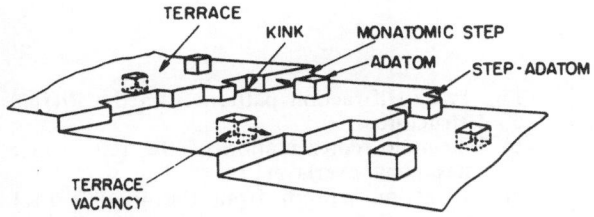

Fig. 6. Model of a heterogeneous solid surface, depicting different sites. These sites are distinguishable by their number of nearest neighbors.

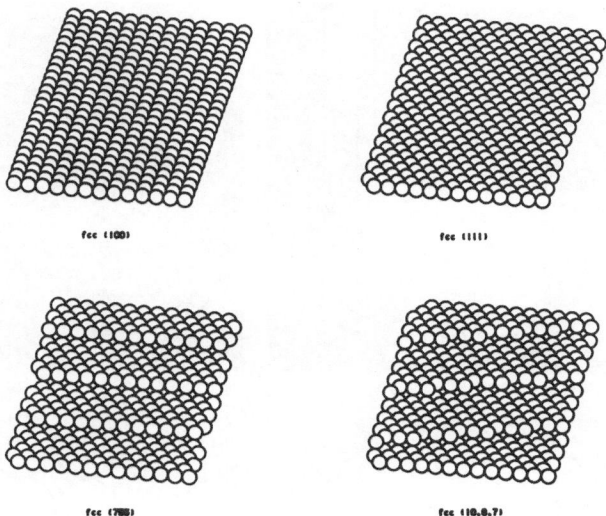

Fig. 7. Idealized atomic surface structures for the flat Pt(111) and Pt(100), the stepped Pt(755), and the kinked Pt(10, 8, 7) surfaces

mixing as shown in figure 5. Clearly, that component that has the lower surface tension or surface free energy is segregated to the surface exponentially in the surface tension difference. It should also be noted that the surface composition is temperature dependent.

2. Structure of clean, solid surfaces

2.1 Surface site geometries

Before we scrutinize the surface chemical bond and surface structure of adsorbed monolayers, we should review what has been learned about the structure of clean, solid surfaces. Figure 6 shows a schematic diagram of a typical surface. Surfaces on an atomic scale are heterogeneous. The various sites are distinguishable by their number of nearest neighbors. Atoms in terraces have the largest number of nearest neighbors; at steps and ledges they have lower coordination.

There is experimental evidence for the presence of all of these surface sites.

Let us concentrate on the structure of these various surfaces, the prototypes of which are shown in figure 7. There are flat surfaces which are low-Miller-index, single-crystal surfaces of cubic materials that have six or four atoms as nearest neighbors in the surface layer. Most surface science studies have been carried out on these flat, low Miller index surfaces. Then there are stepped surfaces which are the typical structure of high Miller index, single-crystal surfaces. Often the steps are also in the direction of high Miller index in which case there are ordered ledges in the steps.

2.2 Surface reconstruction and relaxation

Let us first concentrate on the surfaces of flat, low-Miller-index planes. Figure 8 shows the (100) crystal face of platinum. When clean, this surface is reconstructed and its diffraction pattern indicates the presence of a 5×20 surface structure. When the surface is impure or has a fraction of a monolayer of adsorbates, the square unit cell shown in figure 8 is obtained, which is what one would expect from projection of the bulk unit cell up to the surface. While this 5×20 surface structure was first detected in our laboratory in 1965 [9], it was actually solved by surface crystallography in 1981 [10].

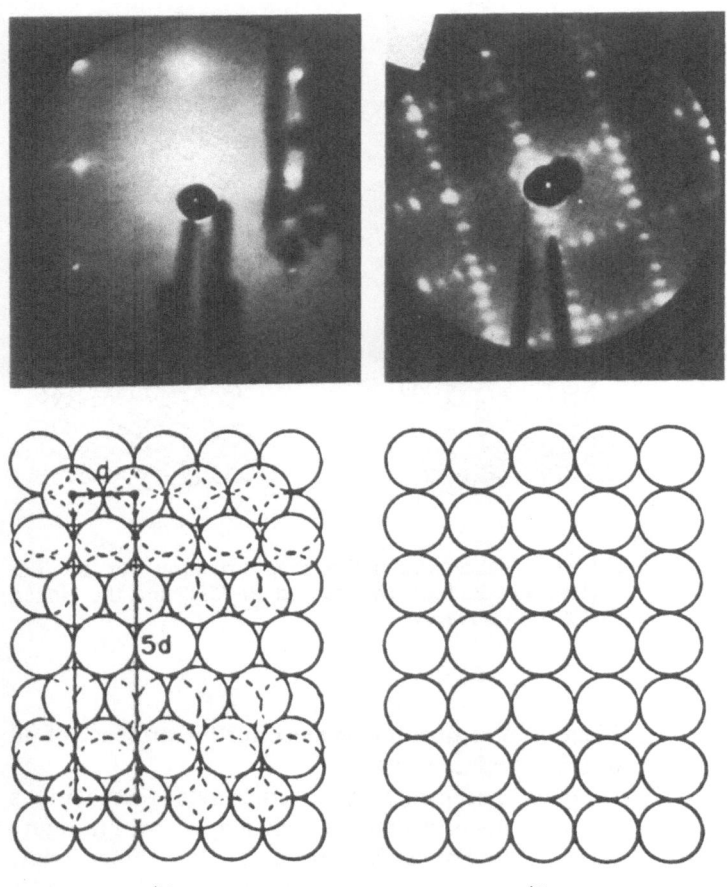

(b)   (d)

Fig. 8. a) Diffraction pattern from the Pt(100) 5 × 1 structure
b) Schematic representation of the 100 surface with hexagonal overlayer
c) Diffraction pattern from the Pt(100) 1 × 1 structure
d) Schematic representation of the 100 surface

fcc (100) : buckled hexagonal top layer

two-bridge   top/center

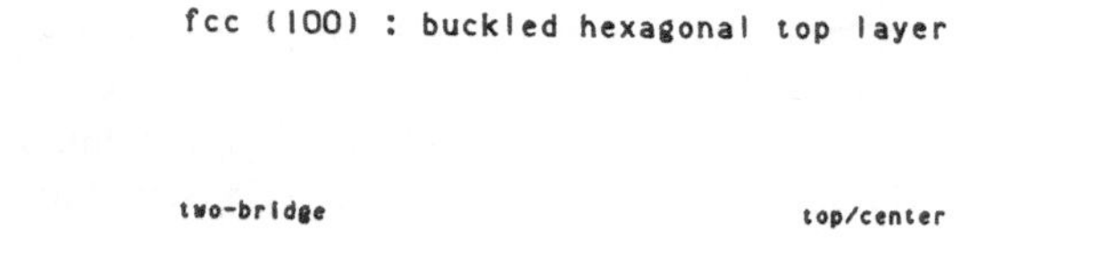

Fig. 9. Structure of the reconstructed Pt(100) crystal face as solved by surface crystallography

This surface structure, with the location of platinum atoms in the surface shown, is depicted in figure 9. The surface is reconstructed into an hexagonal, close-packed structure which sits on top of a square-like second layer. The coincidence of atomic positions between atoms in the first layer and second layer gives rise to the complex diffraction pattern shown in figure 8. This hexagonal reconstruction increases the atomic density in the surface layer which lowers the surface free energy. In addition, it leads to buckling of the surface which increases the total surface energy. A compromise between close-packing and buckling leads to a minimum surface energy which determines the location of atoms in this surface.

Gold, platinum and iridium (100) crystal faces all show reconstruction [1]. Figure 10 shows the stucture of the reconstructed silicon (100) crystal face. In this surface, the silicon atoms form a dimer-like surface structure and there is a relaxation or contraction at

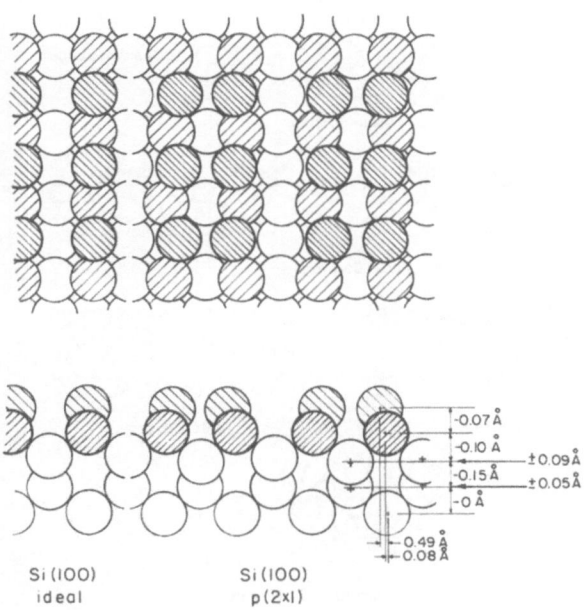

Fig. 10. Top and side views of ideal, bulk-like Si(100) at the left and Si(100) $p(2 \times 1)$ in the modified Schlier-Farnsworth model at the right. Layer-spacing contractions and intra-layer atomic displacement relative to the bulk structure are given. Shading differentiates surface layers

the surface layer with respect to the bulk interlayer distance that extends to three layers below the surface [11]. Thus the effect of the surface is felt three layers into the bulk.

Figure 11 shows the beautiful diffraction patterns exhibited by one of the more stable structures on the reconstructed silicon (111) surface. This is the 7×7 structure which has a complex unit cell which is still not resolved by surface crystallography. It is hoped that a resolution in this surface structure will be forthcoming within a year. It should be noted that the silicon (100) and (111) surfaces are frequently used as substrates for electronic circuitries. As a result, the atomic surface structure of these surfaces is of utmost importance in the integrated circuitry technology, since the electronic transport properties are clearly dependent on the location of atoms at the surface.

There are two major findings of modern surface science that were uncovered during studies of clean, solid surfaces. These are reconstruction, as was demonstrated for the platinum and silicon crystal surfaces, and there is also relaxation. During relaxation the atoms are contracted in their interlayer distance near the surface region with respect to the interlayer distance in the bulk. However, the atomic locations in

Fig. 11. Low-energy electron diffraction patterns taken at four different energies of the reconstructed Si(111) crystal face exhibiting a (7×7) surface structure

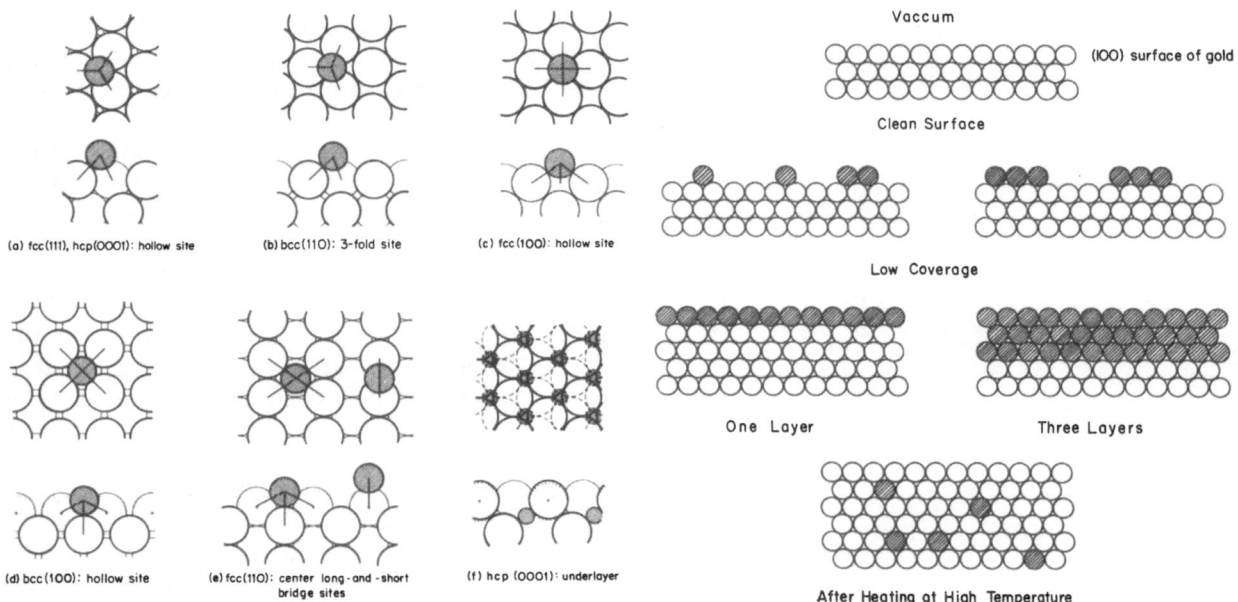

Fig. 12. Top and side views (in top and bottom sketches of each panel) of adsorption geometries on various metal surfaces. Adsorbates are drawn shaded. Dotted lines represent clean surface atomic positions; arrows show atomic displacements due to adsorption

Fig. 13. Scheme of building catalyst monolayers of well-characterized structural composition. Metal atoms are condensed from the vapor phase on single-crystal metal surfaces until desired amounts and atomic structures are obtained

the surface plane (*X, Y* plane) are unchanged. Thus the major conclusion of these clean surface structural studies is that the atomic locations at the surface are different from what one would expect from the projection of the bulk unit cells to the clean surface.

3. The locations of atomic adsorbates on solid surfaces

Over a hundred systems of atomic adsorbates on clean and flat, solid surfaces have been studied. The results indicate that atoms occupy the high symmetry sites, where the next layer of atoms would locate on a growing, single-crystal surface [1]. Some of these atomic positions are shown in figure 12. On surfaces that exhibit hexagonal symmetry, atoms sit in three-fold sites. On surfaces exhibiting square symmetry, atoms sit in four-fold sites.

There are some unique atomic adsorbate bonding situations that should be mentioned. Small atoms such as hydrogen or nitrogen often like to sit below the surface. For example these two atoms sit under the titanium single crystal surface. In the presence of strong chemical interactions there may also be rearrangement of the substrate layer (such is the case for oxygen on the iron (100) crystal face).

Many industrially important systems involve two metals or metals adsorbed on surfaces of other metals. The method of growth in metal deposition on metals is what one would expect from the studies of atomic adsorbates and is shown in figure 13. One can follow such crystal growth with low-energy electron diffraction. An epitaxial relationship seems to be prevalent for metals growing on metal surfaces, that is, the adsorbate interatomic distance seems to mimic the substrate interatomic distance. For example gold, which has a five percent larger interatomic distance than platinum, adapts to the platinum interatomic distance when deposited on (110) single crystal surfaces [12].

Generally, atomic adsorbates seem to covalently bond and sit in high symmetry sites on low Miller index, flat surfaces. This is shown in figure 14, where the atomic distances obtained for adsorbed atoms on single crystal surfaces are compared with atomic distances for solid compounds or for gas phase molecules. The arrows indicate the range of bond lengths that has been observed. This range is clearly within the range of bond lengths observed in both the gas phase and the solid state. Also, the percentage of ionic character of these bonds, as indicated by the charge transfer in figure 14, is very small. Thus, co-

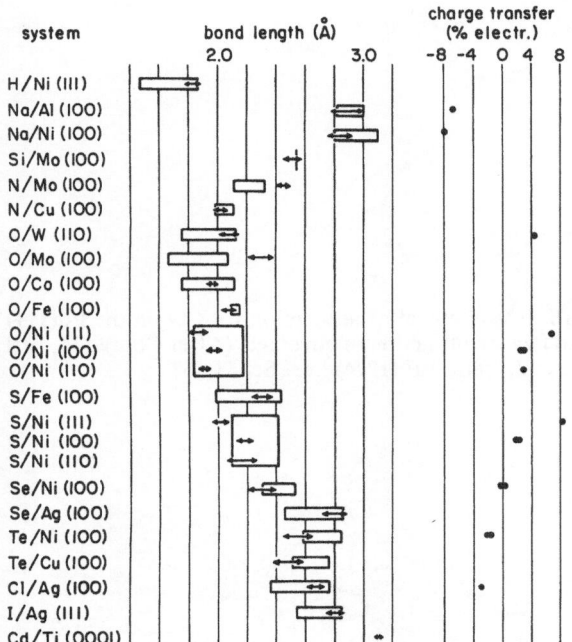

Fig. 14. (left) Comparison of adsorption bond lengths at surfaces (arrows show uncertainty) with equivalent bond lengths in molecules and bulk compounds (blocks extending over range of values found in standard tables). (right) Induced charge transfers for adsorption as determined by work function change, distance of adsorbate from the surface, and surface charge and dipolar charge density

valent bonding predominates for most of the cases that have been studied.

4. The coverage dependence of adsorbate binding at solid surfaces

Thermal desorption spectroscopy provides information about the heat of adsorption of atoms and molecules at solid surfaces [13]. When heating the solid at a well defined rate (in the range of $4-20°$/sec), there is a specific temperature at which the desorption rate is a maximum and from which the heat of desorption can be obtained. When the thermal desorption spectrum is taken at different surface coverages, as shown for carbon monoxide in figure 15, one observes a shift in the peak desorption temperature indicating a variation in the heat of adsorption with coverage.

Such a coverage dependence of the heat of adsorption is shown in figure 16, where the heat of adsorption of carbon monoxide on the palladium (100) surface is shown as a function of surface coverage. At low coverages, the heat of adsorption remains by and

large unchanged. As the coverage reaches about half a monolayer, there is a great drop in the average heat of adsorption per molecule indicating a weakening of the adsorbate substrate bond. This is due to a repulsive interaction between adsorbed carbon monoxide molecules that weakens their bonding to the metal surface.

While in many cases, including adsorbed organic molecules and carbon monoxide, one obtains a repulsive interaction between adsorbed molecules keeping these molecules apart, there are some other cases, including adsorbed hydrogen and oxygen on many metal surfaces, where one obtains an attractive adsorbate-adsorbate interaction that actually increases the average heat of adsorption per molecule at low coverages. Such an attractive interaction also leads to island growth in the adsorbate layer.

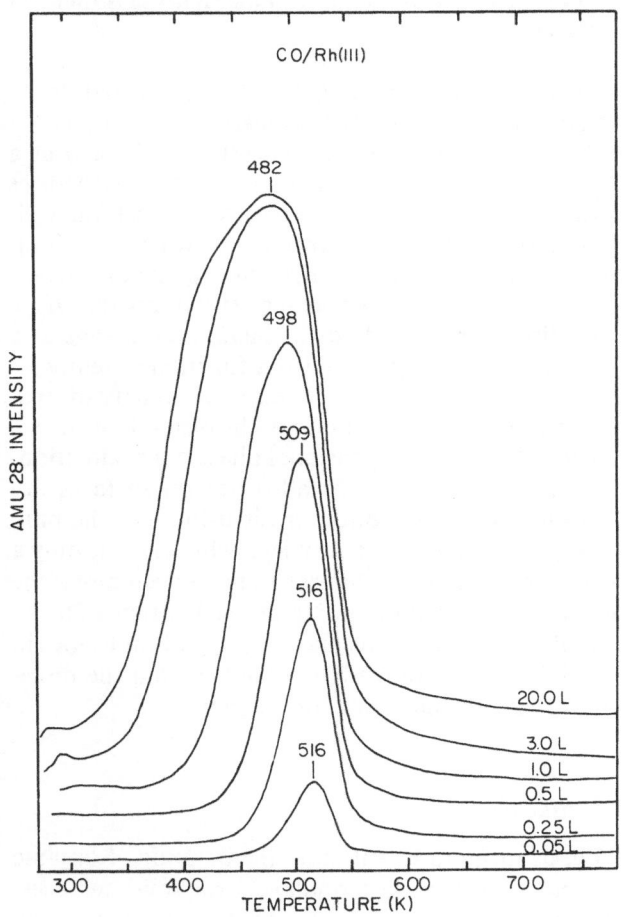

Fig. 15. Thermal desorption spectra of carbon monoxide on Rh(111) measured as a function of coverage following adsorption near 300 K. The crystal heating rate was linear at 15 K/sec. Note the desorption peak temperature shift as a function of coverage

Figures 17 and 18 show the two predominant binding states of carbon monoxide adsorbed on solid surfaces [15, 16]. These are the bridge sites and top sites. Unlike atoms that always occupy the high symmetry sites with three-fold or four-fold coordination, carbon monoxide prefers two-fold or one-fold coordination on most metal surfaces at low coverages. As the coverage of CO is increased, a surface structure with CO in both top and bridge sites, such as that shown in figure 19, may form. In this case, because of repulsive interaction between molecules, the molecules in top sites move sideways to occupy a pseudo hexagonal structure site that is most stable at these high coverages. Here we notice a balance of the adsorbate-substrate and adsorbate-adsorbate interaction that clearly controls the location of molecules on surfaces.

5. The structure of small, organic molecules on solid surfaces

Surface crystallography, as a result of rapid development of the theory of low-energy electron diffraction and low-energy electron scattering, is now at a point where it can solve complex surface structures with large unit cells with many molecules per unit cell. Several of these will be shown below when we exhibit the surface structures of different organic molecules. There are many ordered surface structures that have been discovered, hundreds of them, that change as a function of coverage as well as a function of temperature. The reason for the richness and diversity of two-dimensional surface structural chemistry is what we would call the two dimensional phase approximation. The molecules and atoms adsorbed on surfaces are protected from desorption or diffusion into the bulk by large potential energy barriers, while there is only a small potential energy barrier to movement along the surface. This situation is diagrammed in figure 20. As a result, rearrangements and ordering of molecules in two dimensions are readily possible within the molecules' long residence time on surfaces.

5.1 Alkenes

Let us turn our attention to the bonding of organic molecules in organic monolayers on solid surfaces. The first molecule whose surface structure was solved was ethylene on flat metal surfaces such as platinum (111) and rhodium (111) [18, 19]. The structure of the chemisorbed ethylene molecule at room temperature is shown in figure 21. Ethylene loses a hydrogen, be-

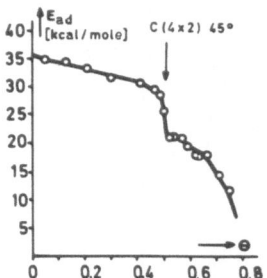

Fig. 16. Isosteric heat of adsorption for CO on the Pd(111) crystal face as a function of coverage. (After Conrad H, Ertl G, Koch J, Latta EE (1974) Surf Sci 43:462

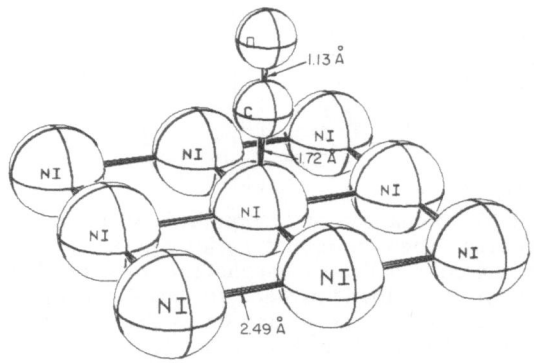

Ni (100) + c (2×2) CO

Fig. 17. Top site bonding structure of carbon monoxide on Ni(100) from low energy electron diffraction and electron spectroscopy studies

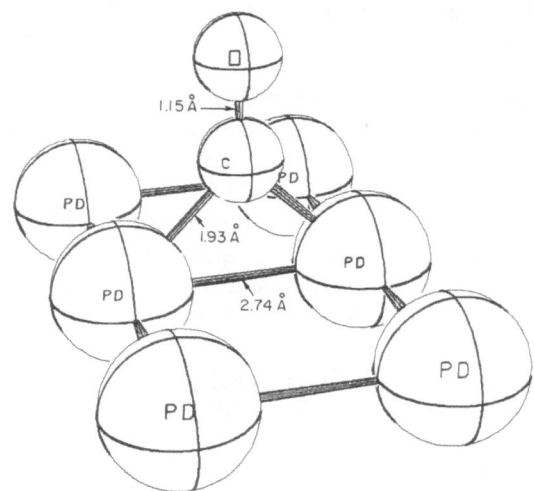

Pd (100) + (2√2×√2) R 45° 2 CO

Fig. 18. Bridge site adsorption structure of carbon monoxide on Pd(100) at a half monolayer coverage

Rh(III) (2×2) - 3 CO

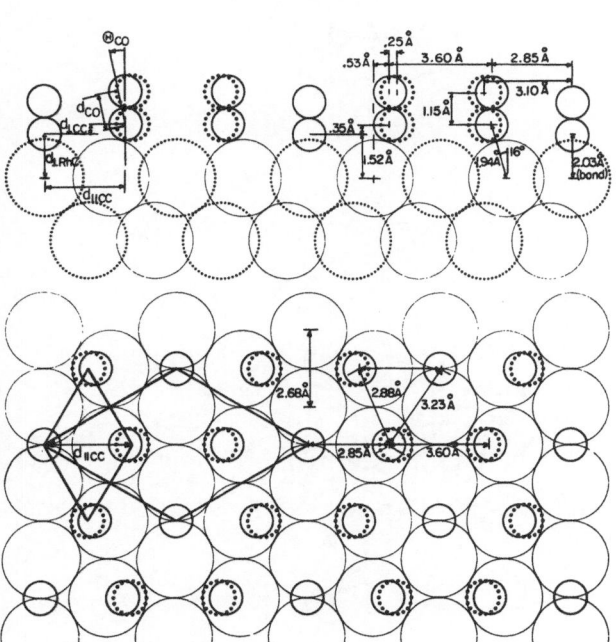

Fig. 19. Structure determined by low energy electron diffraction for a saturation coverage of carbon monoxide on Rh(111). Top and side views are shown. Large circles represent Rh atoms while smaller circles correspond to C and O atoms. Solid lines show the structure expected for hexagonal close-packing of the carbon monoxide while dotted circles depict the actual structure

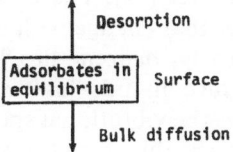

Exceptions: low pressure-high surface temperature studies (molecular beam-surface scattering) exothermic surface reactions?

Fig. 20. The two-dimensional phase approximation. Potential energy barriers for desorption or bulk diffusion are much larger than for surface diffusion, so equilibrium is attained in two dimensions only

coming C_2H_3, and rearranges to bond in a three-fold hollow as shown. This species is known as ethylidyne. The carbon atom closest to the surface has a metal-carbon distance of 2.0 Å, shorter than the 2.2 Å covalent metal-carbon bond predicted using covalent

Different ethylidyne species: bond distances and angles
(r_C = carbon covalent radius; r_M = bulk metal atomic radius)

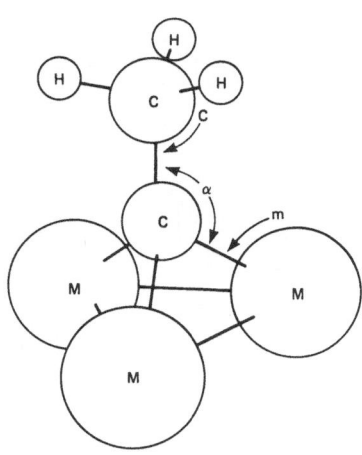

	C [Å]	m	r_M	r_C	α [°]
$Co_3 (CO)_9$ CCH_3	1.53 (3)	1.90 (2)	1.25	0.65	131.3
$H_3 Ru_3 (CO)_9$ CCH_3	1.51 (2)	2.08 (1)	1.34	0.74	128.1
$H_3 Os_3 (CO)_9$ CCH_3	1.51 (2)	2.08 (1)	1.35	0.73	128.1
Pt (111) + (2 × 2) CCH_3	1.50	2.00	1.39	0.61	127.0
Rh (111) + (2 × 2) CCH_3	1.45 (10)	2.03 (7)	1.34	0.69	130.2
$H_3C - CH_3$	1.54			0.77	109.5
$H_2C = CH_2$	1.33			0.68	122.3
$HC \equiv CH$	1.20			0.60	180.0

Fig. 21. The surface structure of ethylidyne (bond distances and angles) is compared with several tri-nuclear metal cluster compounds of similar structure

radii of the atoms. The carbon-carbon bond is perpendicular to the surface and is stretched to a single bond length.

The same structure is proven by parallel high-resolution electron energy loss spectroscopy studies, and figure 22 shows the vibrational spectrum of adsorbed ethylidyne molecules on the rhodium (111) crystal face [20]. The C_{3v} symmetry determined from peak intensities and the vibrational frequencies clearly coincide with the structure obtained by surface crystallography. The remarkable agreement between the vibrational frequencies for ethylidyne on the surface and ethylidyne in a organometallic, tri-metal cluster $[(CCH_3)Co_3(CO)_9]$ indicates that a localized bonding model can accurately describe the surface bonding of organic fragments [21].

Figure 23 shows the similar alkylidyne structures of chemisorbed propylene and butenes that have also

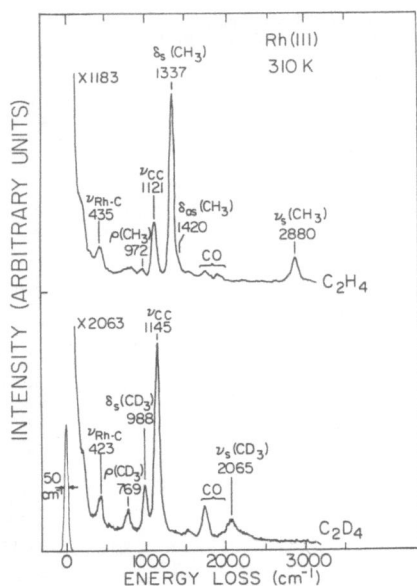

Fig. 22. High-resolution electron energy loss vibrational spectrum of ethylidyne (CCH₃) and ethylidyne-d_3 on Rh(111): the stable, room temperature, chemisorbed structure for ethylene

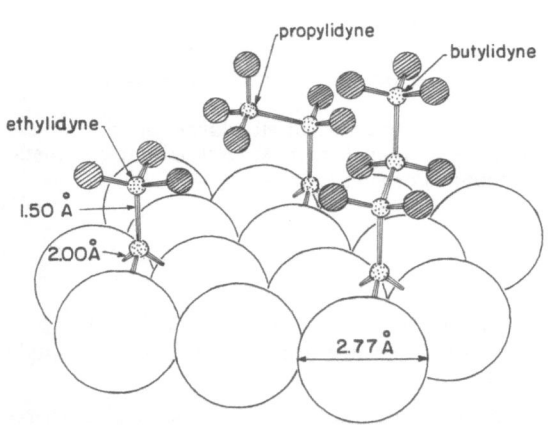

Pt (III) + ethylidyne, propylidyne and butylidyne

Fig. 23. Surface structures for alkylidyne species formed on Pt(111) after the adsorption and rearrangement of ethylene, propylene, and butenes. These structures were determined by surface crystallography and high resolution electron energy loss spectroscopy

been found to exist on the (111) crystal faces of platinum, rhodium and palladium. It appears that alkylidyne structures predominate in the bonding of small alkenes on transition metal surfaces at room temperature.

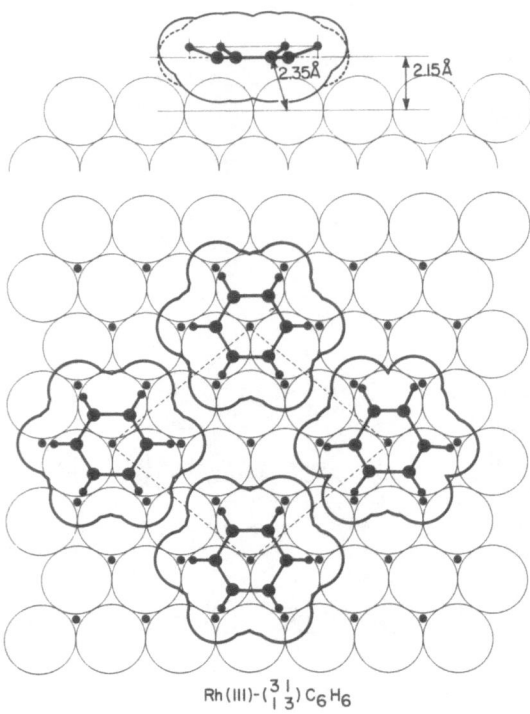

$Rh(III)-\binom{3\ 1}{1\ 3} C_6 H_6$

Fig. 24. Surface structure of benzene as determined from low-energy electron diffraction studies and surface crystallography

5.2 Benzene

Figure 24 shows the surface structure of benzene of the rhodium (111) crystal face [22]. There are several ordered surface structures that change with coverage. This is one of those. Clearly, benzene lies with its π ring parallel to the surface in this high symmetry structure. Figure 25 shows the vibrational spectrum of benzene chemisorbed with this surface structure. Again, the high symmetry structure is proven by at least two techniques — low-energy electron diffraction and high-resolution electron loss spectroscopy.

5.3 Alkanes

Alkanes may be associatively adsorbed on metal surfaces. Figure 26 shows the ordering and structures of alkanes deposited on metal surfaces as a function of temperature [23]. As the adsorption temperature is decreased, these molecules form first a disordered, then an ordered monolayer. As the temperature is lowered still further in the presence of the organic vapor, condensation occurs leading to crystal growth of an organic single crystal. In this way, not only ad-

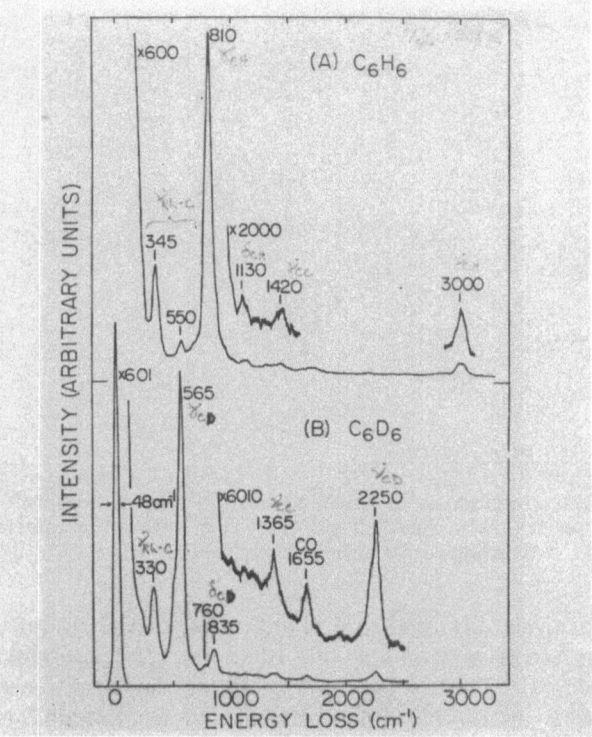

Fig. 25. The vibrational spectra of benzene and deuterated benzene adsorbed on a Rh(111) surface as determined by high-resolution electron energy loss spectroscopy

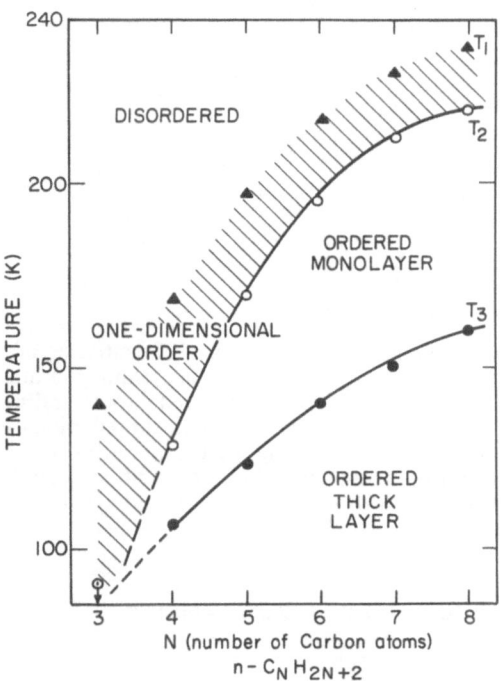

Fig. 26. Monolayer and multilayer surface phases of the n-paraffins $C_3 - C_8$ on Pt(111) and the temperatures at which they are observed at 10^{-7} Torr

sorbed monolayers, but also growing crystals of adsorbates can be studied by surface crystallography.

6. Coadsorption of atoms and molecules on solid surfaces

6.1 Site-blocking: $S + H_2$ on Mo(100)

It is frequently important to study the presence of two or more atoms or molecules that are simultaneously adsorbed on solid surfaces. There are some cases where the adsorption of one atom blocks the adsorption of the other one. This is the case for the adsorption of hydrogen on a molybdenum (100) surface that is partially covered with sulfur as shown in figure 27 [24]. Here the binding of hydrogen is unaffected by the coadsorption of sulfur. However, the amount of hydrogen that is adsorbed declines rapidly with sulfur coverage. This rapid decline in hydrogen adsorption occurs because every sulfur atom removes the adsorp-

tion possibility for two hydrogen atoms which require unoccupied, adjacent sites for adsorption.

6.2 Electronic interaction: $K + CO$ on Pt(111) and Rh(111)

Interaction between adsorbates is much stronger in the coadsorption of potassium atoms and carbon monoxide on transition metal surfaces. When potassium is adsorbed on a transition metal surface at low coverages, it is completely ionized by charge transfer to the transition metal. When carbon monoxide is coadsorbed with potassium [14], the bonding of carbon monoxide to the metal is substantially strengthened through the strengthening of the metal-carbon bond, while the carbon oxygen bond is substantially weakened, increasing the probability for the dissociation of carbon monoxide.

This bonding change is exhibited in figure 28 where the vibrational spectrum of carbon monoxide on Pt(111) with increasing concentration of potassium is shown [25]. This figure clearly shows the weakening of the carbon-oxygen bond by the shift in the CO stretching frequency to lower energy. The bond-

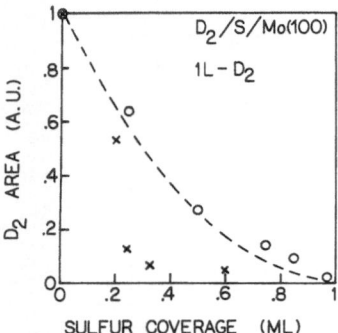

Fig. 27. The site-blocking effect of sulfur on deuterium adsorption on Mo(100) as determined by deuterium thermal desorption. ($\bigcirc$) = sulfur layer disordered, ($\times$) = sulfur layer ordered. The broken line is a theoretical prediction of the site-blocking effect assuming that one sulfur atom blocks one deuterium atom adsorption site and that deuterium molecules chemisorb dissociatively in adjacent, unoccupied sites

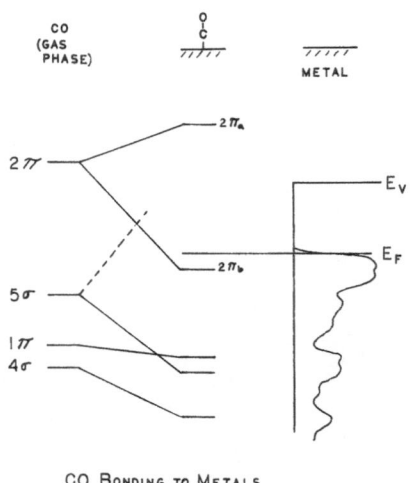

Fig. 29. Interaction of gas phase carbon monoxide molecular orbitals with the filled energy levels of a transition metal to form bonding orbitals for associatively adsorbed CO

weakening amounts to a change from a CO double bond to a one and a half bond on this platinum surface. This electronic interaction between adsorbates can be explained by the molecular orbital diagram shown in figure 29. Charges from the potassium change the density of states of electrons at the Fermi level in the transition metal. These electrons from the transition metal in turn find their way to antibonding and bonding molecular orbitals of the coadsorbed organic molecules, in this case carbon monoxide.

Interestingly, such interaction can lead to complete dissociation of the carbon monoxide molecule, as shown by isotope scrambling experiments on Rh(111)

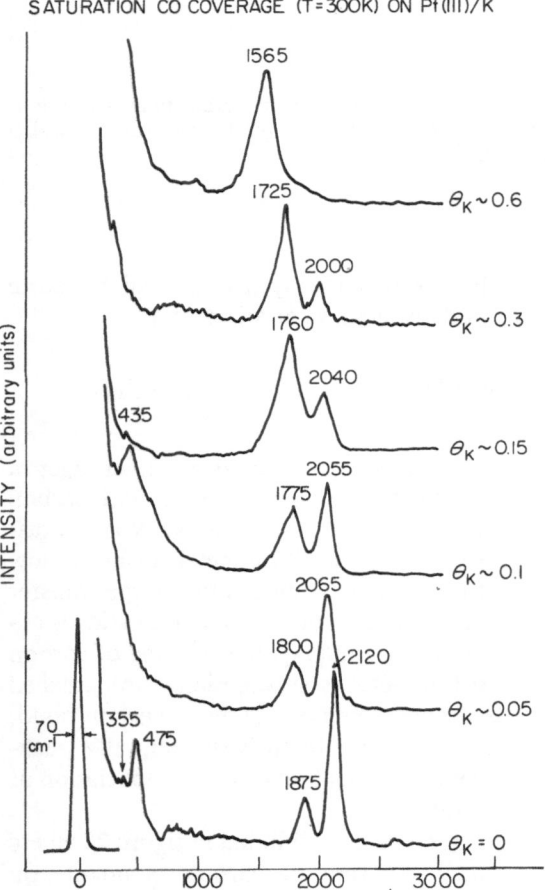

Fig. 28. Vibrational spectra of the saturation carbon monoxide coverage chemisorbed on Pt(111) at 300 K as a function of preadsorbed potassium coverage

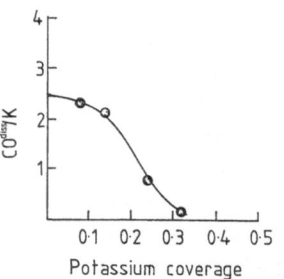

Plot of number of CO molecules that dissociate per potassium as a function of potassium coverage.

Fig. 30. Carbon monoxide dissociation on a Rh(111) surface as a function of potassium coverage as determined by thermal desorption isotope scrambling experiments with $C^{13}O^{16}$ and $C^{12}O^{18}$

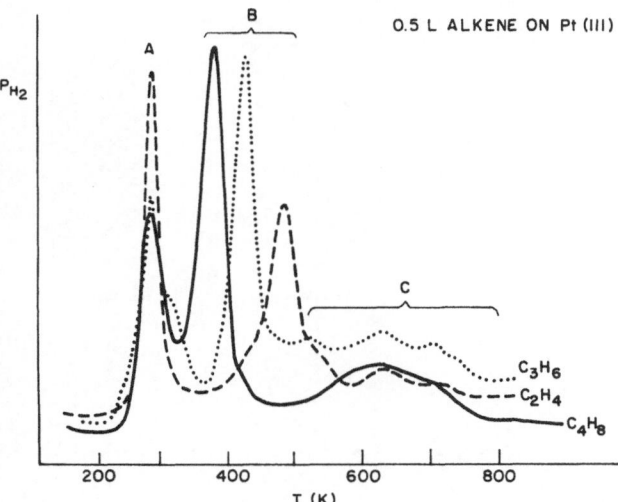

Fig. 31. Hydrogen thermal desorption spectra illustrating the sequential dehydrogenation of ethylene, propylene, and cis-2-butene chemisorbed on the Pt(111) crystal surface at 120 K. The rate of heating is 12 K/sec

7. Temperature dependent changes of bonding of adsorbed molecules on solid surfaces

When an organic molecule is adsorbed on a transition metal surface and then heated, sequential dehydrogenation instead of desorption is observed. This is shown in figure 31. From alkenes, hydrogen evolution is observed sequentially at well defined temperatures indicating that organic fragments are left behind on a transition metal surface. These organic fragments have been studied by a variety of techniques, the most powerful of which appears to be high-resolution electron energy loss spectroscopy. Figure 32 shows the fragments that have been identified so far. There are CH, C_2, CH_2, and C_2H fragments that are all detectable. It appears that while these fragments would be free radical like in the gas phase, due to the strong metal carbon interaction, these are highly stable and well characterizable in well-defined temperature ranges on surfaces [18]. Only at the highest temperature of heat treatment does all the hydrogen release and the surface carbon graphitize to reach the thermodynamic end product of such a metal-organic molecule interaction.

Thus, there is sequential bond-breaking in adsorbed species which leads to the formation of surface intermediates that may be metastable in the thermodynamic sense. However, they are very stable in a well defined temperature range. In fact, one can adsorb on a transition metal surface any number of very reactive organic molecules, and if the temperature is low enough, there is no chemical reaction. As the temperature is increased, there is sequential bond breaking and a CH, CC, CO, or CN bond will break leaving a fragment which is stable in a finite temperature range. Beyond this temperature, again, another bond

[26]. By exposing an alkali predosed surface to two carbon monoxide labeled isotopes ($C^{13}O^{16}$ and $C^{12}O^{18}$), and looking for the scrambling of these two in the thermal desorption products ($C^{13}O^{18}$, amu = 31), one can prove that carbon monoxide dissociation occurs. This happens of course only in the presence of a potassium adlayer. In the absence of such a layer, carbon monoxide does not dissociate on this clean transition metal surface at low pressures. Figure 30 shows the number of carbon monoxide molecules that dissociate per potassium atom. This number can reach a value greater than two at low potassium coverages.

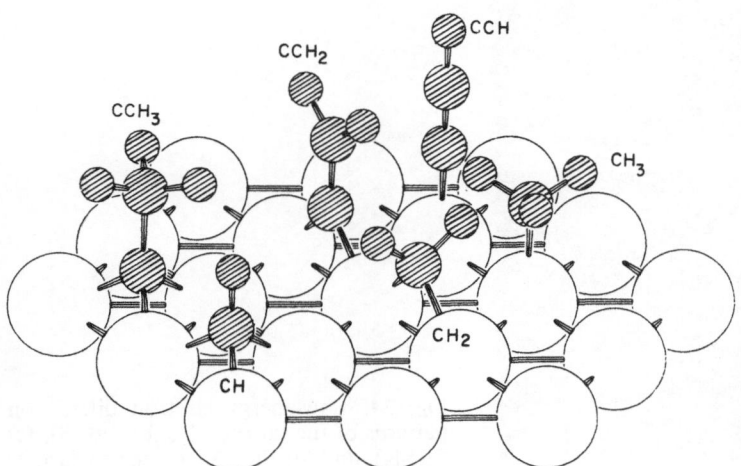

Fig. 32. Schematic representation of the various organic fragments that are present on metal surfaces at higher temperature. The presence of CH, C_2, C_2H, CH_2, and CCH_3 species has been detected

breaking process occurs and again another fragment is left on the surface. Only at the very highest temperature will thermodynamically stable species form in the case of organic adsorbates. This stable state is a graphitized surface and hydrogen in the gas phase.

8. Adsorption and chemical bonding on high Miller index, stepped and kinked surfaces

Figure 33 shows the surface structures of several stepped surfaces of high Miller index. These surfaces are stable in a stepped/terrace configuration [26]. On the clean surface of a close-packed metal, the steps are usually of one atom in height, periodically distributed, and are separated by terraces of roughly equal width. Typical diffraction patterns of such surfaces are shown in figure 34. The formation of doublets or triplets indicates the appearance of new periodicities from which the stepped structure of these high Miller index surfaces can be obtained.

When atoms or molecules adsorb on these high Miller index surfaces, they have available to them now several additional sites where their binding could be different. This is clearly indicated by thermal desorption spectroscopy studies. Figure 35 shows that carbon monoxide exhibits two desorption peaks at high coverages while at low coverages only the higher temperature desorption peak is present. The two desorption peaks, when compared with flat surfaces, can be attributed to carbon monoxide adsorbed at stepped

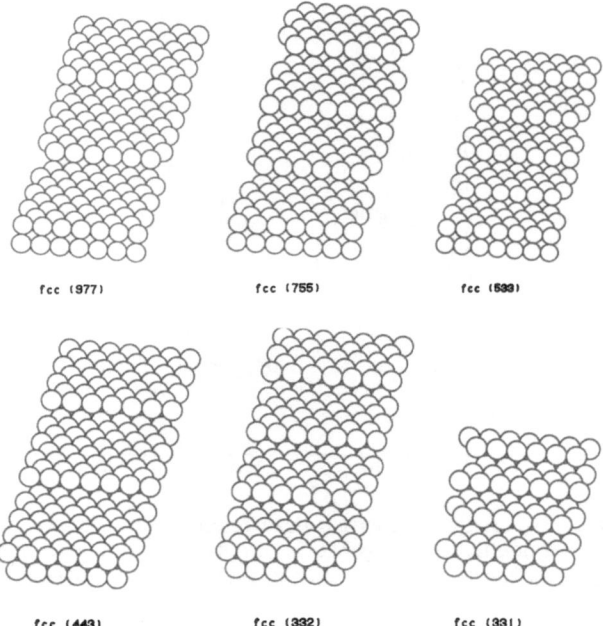

Fig. 33. Structure of several high Miller index stepped surfaces with different terrace widths and step orientations

sites as well as the terrace sites. Since the adsorbed molecules at stepped sites desorb at the higher temperature, their binding is stronger than that at the terrace sites.

Figure 36 shows the desorption peaks of hydrogen after adsorption on flat, stepped and kinked platinum

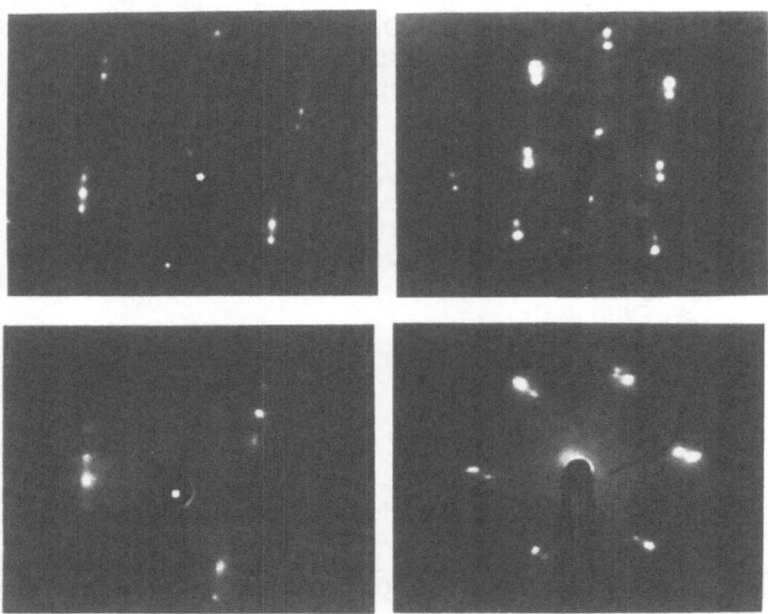

Fig. 34. Low-energy electron diffraction patterns of the (a) Pt(755), (b) Pt(679), (c) Pt(544), and (d) Pt(533) stepped surfaces

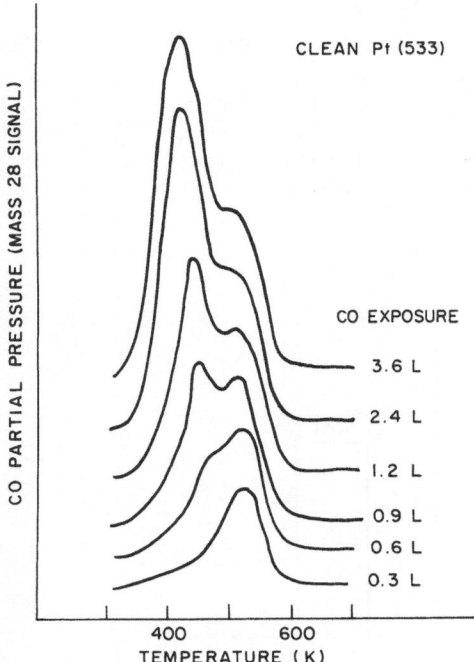

Fig. 35. Thermal desorption spectra of carbon monoxide from a Pt(533), stepped crystal face as a function of coverage. The two peaks are indicative of CO bonding at step and terrace sites. The higher temperature peak corresponds to CO bound at step sites

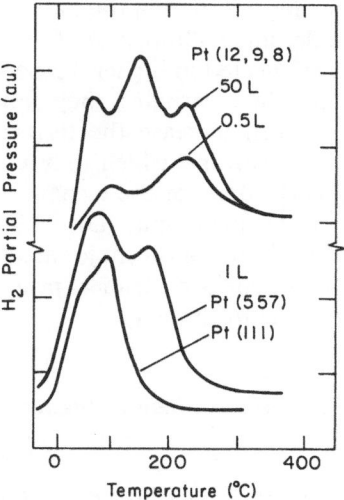

Fig. 36. Thermal desorption spectra for hydrogen chemisorbed on flat Pt(111), stepped Pt(557), and kinked Pt(12, 9, 8) surfaces

surfaces. On the kinked surfaces there are three desorption peaks. On the stepped surfaces there are two and on the flat surfaces there is only one. These desorption spectra make it relatively easy to associate the highest temperature desorption peak with hydrogen atoms at kinked sites, the middle one with hydrogen atoms at stepped sites, and the lowest temperature desorption peak with hydrogen atoms on terrace sites.

The surface chemical bond is highly structure sensitive. This is the major conclusion of studies of single crystal surfaces with adsorbates. Thus, if one measures the heat of adsorption as a function of atomic (Z) number across a transition metal series, one finds a diversity and richness of chemical bonding that is exhibited in figure 37. Clearly, on each surface there may be many binding sites where the adsorbate is bound with different binding energies. There is no such thing as one adsorbate-substrate chemical bond. There may be four or five different bonding geometries with different heats of adsorption for a given atom or molecule on a given surface. This is one of the major reasons for the richness and diversity of surface chemistry that is exhibited in heterogeneous catalysis.

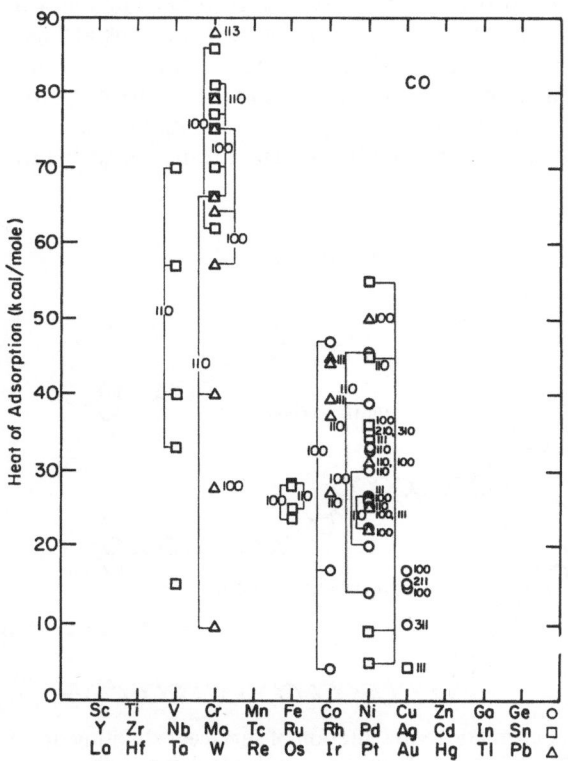

Fig. 37. Heats of adsorption of carbon monoxide on single crystal surfaces of transition metals

One should not give the impression that the stepped surfaces remain stable under all conditions. As figure 38 shows, when surfaces are heated to higher temperature in the presence of certain adsorbates, they may undergo rearrangements which increase the terrace width as well as the step height and which may ultimately lead to faceting [27]. Many of these changes may be reversible; many of them may not. The stability of surface structures is one of the major areas of surface science, since many surface chemical properties depend on surface structure stability.

9. The effect of surface structure in heterogeneous catalysis

Surface structure sensitivity is well-illustrated by example from the field of catalysis. Figure 39 shows the rate of formation of ammonia from nitrogen and hydrogen on iron single crystal surfaces [28]. The open (111) crystal face is about 500 times more active in ammonia production than the close-packed (110) crystal face. In this reaction the dissociation of dinitrogen to nitrogen atoms is a rate-determining step. On the (111) surface there are sites where this dissociation process can occur with near zero activation energy. In these dissociation reactions, atoms at high coordination sites, which are found in the second layer near the surface, are implicated as the sites for bond-breaking. A more open surface makes the second layer high coordination metal atoms accessible to

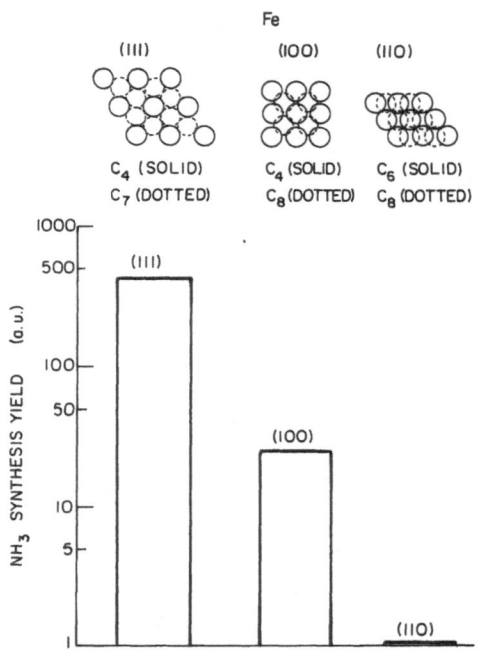

Fig. 39. The remarkable surface structure sensitivity of the iron-catalyzed, ammonia synthesis

the incoming adsorbates, and this is one reason for the great reactivity of these more open surfaces.

The importance of surface structure in reactions can be shown by another example, i.e. the hydrogen-deuterium exchange using mixed molecular beams of these two isotopes [29]. Figure 40 shows the scheme of the experiments. An incoming mixed molecular beam of hydrogen and deuterium is incident on a single crystal surface, and the product distribution is monitored as a function of angle with a mass spectrometer. By chopping the scattered beam, the velocity of the scattered beam can also be determined.

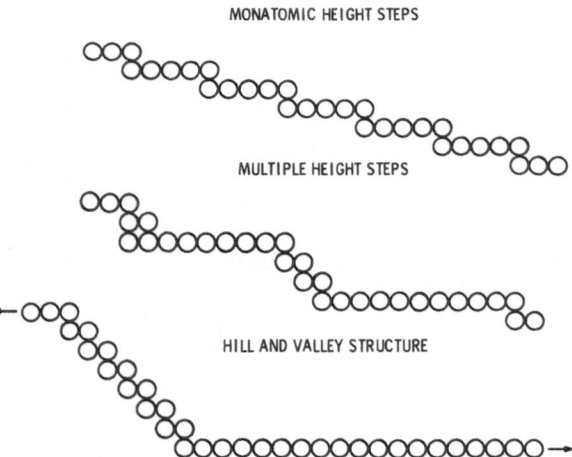

Fig. 38. Schematic representation of surfaces exhibiting one-atom step height configuration, multiple-height step structure, and hill-and-valley configuration consisting of large facet planes. Reconstruction from one type to another may occur on adsorption and/or heating

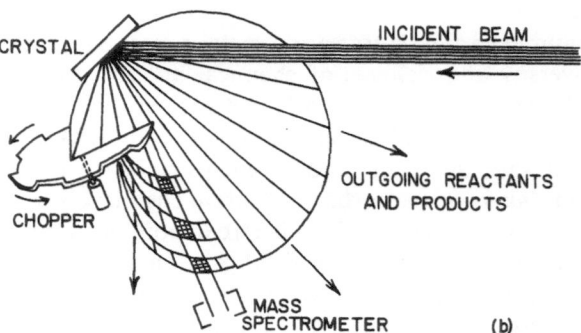

Fig. 40. Scheme of the molecular beam — surface scattering experiment

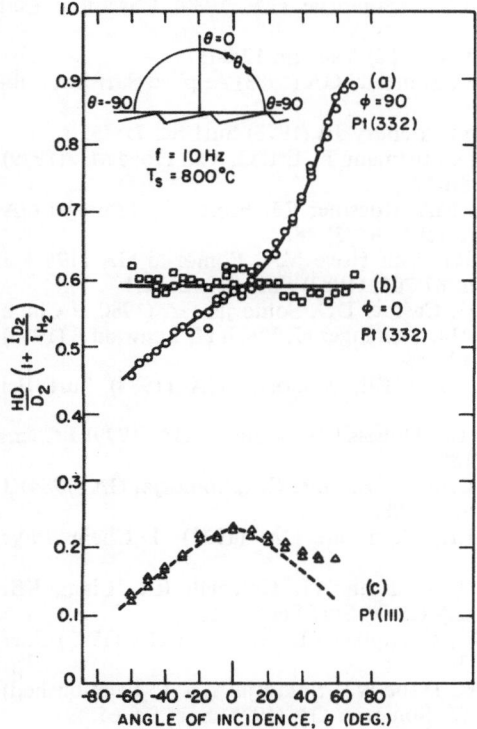

Fig. 41. HD production as a function of angle of incidence, θ, of the molecular beam, normalized to the incident D_2 intensity. (a) Pt(332) with step edges perpendicular to the incident beam ($\Phi = 90°$); (b) Pt(332) where the projection of the beam on the surface is parallel to the step edges ($\Phi = 0°$); (c) Pt(111)

Figure 41 shows the reaction probability of forming HD from H_2 and D_2 on a stepped surface when the mixed beam hits the open side of the steps so that the fraction of atoms at the bottom of the step exposed is almost unity. As the beam changes direction so that the bottom of the steps are no longer exposed but are shielded, the reaction probability drops by a factor of 2. On the (111) surface where there are no steps the reaction probabiltiy is again down by about an order of magnitude. Clearly, in this case, H–H bond breaking occurs and this process is also highly structure sensitive.

10. Application of modern surface science techniques to study the solid-liquid interface

Recently a new technology was developed in our laboratory which is now applied widely in a large number of research groups that combines ultra-high vacuum surface science with high pressure catalytic reaction studies [30]. The cell that is used for this purpose is shown in figure 42. The surface which is first exposed to ultra-high vacuum and characterized to determine the surface structure and surface composition is then isolated by an isolation cell which can be pressurized to several atmospheres to carry out a reaction study. Then the cell can be pumped, evacuated and opened, and the sample is sitting again in ultra-high vacuum where surface science studies can determine changes of surface structure and composition caused by the high pressure catytic reaction.

Such a cell is also applicable for studies of reactions at solid-liquid interfaces that are important in electrochemistry and in colloid chemistry. In electrochemistry, several cells that can be filled with liquids to study electrode reactions have been reported [31]. After electrochemical studies, the liquid is removed and the electrode is studied by modern surface science techniques.

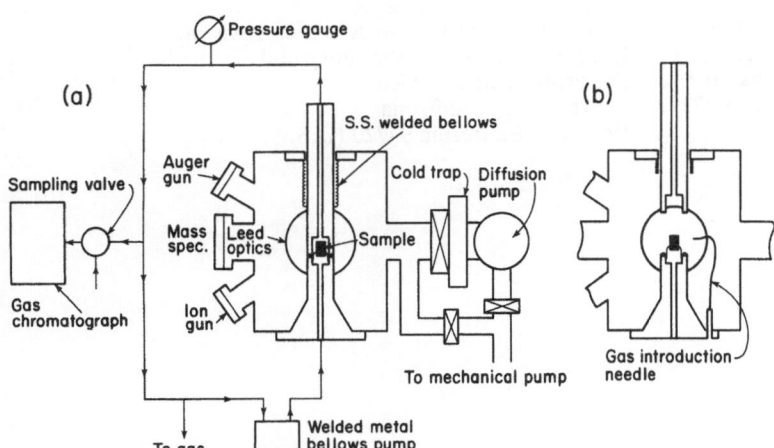

Fig. 42. Schematic representation of the experimental apparatus to carry out catalytic-reaction-rate studies on single crystal surfaces of low surface area at low and high pressures in the range of 10^{-7} to 10^4 Torr

We believe this approach will go a long way in allowing a molecular scrutiny of the solid-liquid interfaces that are present in colloid systems. We believe this is an important direction for surface science that will be a great asset to colloid scientists and those scientists interested in the molecular ingredients of systems and reactions at the solid-liquid interface.

Acknowledgements

This work was supported by the Director, Office of Energy Research, Office of Basic Energy Sciences, Chemical and Materials Sciences Division of the U.S. Department of Energy under Contract Number DE-AC03-76SF00098.

References

1. Somorjai GA (1981) Chemistry in Two Dimensions: Surfaces. Cornell University Press
2. Ertl G, Küppers J (1979) Low Energy Electrons and Surface Chemistry. Verlag Chemie, Weinheim
3. Van Hove MA, Tong SY (1979) Surface Crystallography by Low-Energy Electron Diffraction: Theory, Computation and Structural Results. Springer-Verlag, Heidelberg
4. Ibach H, Mills DL (1982) Electron Energy Loss Spectroscopy and Surface Vibrations. Academic Press, New York
5. Feuerbacher B, Fitton B, Willis RF (eds) (1979) Photoemission and the Electronic Properties of Surfaces. Wiley, London
6. Chang CC (1974) Analytical Auger Electron Spectroscopy. In: Kane PF, Larrabee GB (eds) Characterization of Solid Surfaces. Plenum Press, New York
7. Jablonski A, Overbury SH, Somorjai GA (1977) Surf Sci 65:578
8. Overbury S, Bertrand P, Somorjai GA (1975) Chem Rev 75 (5):550
9. Hagstrom S, Lyon HB, Somorjai GA (1965) Phys Rev Lett 15:491
10. Van Hove MA, Koestner RJ, Stair PC, Biberian JP, Kesmodel LL, Somorjai GA (1981) Surf Sci 103:189, 218
11. Appelbaum J, Hamann DR (1978) 74:21; Mitchell KAR, Van Hove MA (1978) Surf Sci 75:147L; Tong SY, Maldonado AL (1978) Surf Sci 78:459; Appelbaum JA, Baraff GA, Hamann DR (1975) Phys Rev Lett 35:729; (1976) Phys Rev B14:588
12. Sachtler JWA, Somorjai GA (1980) Phys Rev Lett 45:1601
13. Redhead PA (1962) Vacuum 12:203
14. Crowell JE, Somorjai GA (1983) Applied Surf Sci, Submitted Nov
15. Andersson S, Pendry JB (1978) Surf Sci 71:75
16. Behm HJ, Christmann K, Ertl G, Van Hove MA (1979) Surf Sci 88:L59
17. Van Hove MA, Koestner RJ, Frost JC, Somorjai GA (1983) Surf Sci 129(2/3):482
18. Koestner RJ, Van Hove MA, Somorjai GA (1983) J Phys Chem 87:203; (1982) Surf Sci 121:321
19. Dubois LH, Castner DG, Somorjai GA (1980) J Chem Phys 72:5234; Steininger H, Ibach H, Lehwald S (1980) Surf Sci 117:685
20. Koel BE, Bent BE, Somorjai GA (1984) Surf Sci 146:211
21. Kesmodel LL, Dubois LH, Somorjai GA (1979) J Chem Phys 70:2180
22. Koel BE, Crowell JE, Mate CM, Somorjai GA (1984) J Phys Chem 88:1988
23. Firment LE, Somorjai GA (1977) J Chem Phys 66(7):2901
24. Farias MH, Gellman AJ, Chianelli RR, Liang KS, Somorjai GA (1984) Surf Sci 140:181
25. Crowell JE, Garfunkel EL, Somorjai GA (1982) Surf Sci 121:303
26. Crowell JE, Tysoe WT, Somorjai GA (to be published)
27. Blakely DW, Somorjai GA (1977) Surf Sci 65:491
28. Spencer ND, Schoonmaker RC, Somorjai GA (1982) J Catal 74:129; (1981) Nature 294:643
29. Salmeron M, Gale RJ, Somorjai GA (1979) J Chem Phys 70(06):2807
30. Cabrera AL, Spencer ND, Kozak E, Davies PW, Somorjai GA (1982) Rev Sci Instrum 53:1888; Blakely DW, Kozak EI, Sexton BA, Somorjai GA (1976) J Vac Sci Technol 13:1091
31. Hubbard AT (1980) Acc Chem Res 13:177; Ross PN Jr (1982) In: Vanselow R, Howe R (eds) Chemistry and Physics of Solid Surfaces IV. Springer-Verlag

Received July 18, 1984;
accepted July 30, 1984

Authors' address:

Professor G. A. Somorjai
Materials and Molecular Research Division
Lawrence Berkeley Laboratory and
Department of Chemistry
University of California
Berkeley, California 94720 (U.S.A)

Progress in Colloid & Polymer Science .   Progr Colloid & Polymer Sci 70:57 – 61 (1985)

Non-local electronic perturbations of metal-carbonyl bonds

J. Paul

Department of Physics, Chalmers University of Technology, Göteborg (Sweden)

Abstract: Metal-carbonyl bonds, as found for carbon monoxide either liganded to metalloporphyrins or chemisorbed onto metal surfaces, are discussed. These two classes of systems are compared with emphasis on short range vs. long range perturbations of the vibrational bands, ν_{MeC} and ν_{CO}. It is concluded that metalloporphyrins serve as models for terminal bonding on metal surfaces and that data obtained for carbonyl-hemes enable the isolation of non-local electronic effects, which are much discussed in connection with alkali promotors co-adsorbed with CO on metal catalysts.

Key words: Metal-carbonyl bonds, non-local electronic perturbations, carbonyl-hemes, chemisorbed carbon monoxide

1. Introduction

Metal atoms act as reactive centres in various chemical environments. Catalysts in the form of metal surfaces or zeolites determine processes in man-made chemical reactors. In biological systems metalloproteins direct reactions of living cells with high selectivity. Some elements operate in a particular reaction, but specificity and even efficiency can be markedly influenced by a synthetic or natural support material. A few adsorbed potassium atoms will turn a relatively inert iron surface into an efficient cracking agent for carbon monoxide. It is even more spectacular that only apparently small geometrical differences exist between the immediate surroundings of iron in various hemeproteins in spite of the fact that such proteins mediate a large variety of reactions. Ironporphyrins, hemes are the prosthetic groups in hemeproteins.

It is generally accepted that intermediate complexes between metal atoms and reactants play a crucial role in most metal-catalyzed reactions. Consequently to understand more completely such reactions on the atomic level one should focus on the response of metal electronic structures to perturbations in the vicinity of the active, binding site. Such perturbations can be studied in several ways. One way is to probe intermediate states directly be means of some sensitive and relevant spectroscopy. A second way is to use other molecules as *in situ* monitors of binding sites. A third way is to make use of electron structure calculations.

The vibrational bands of metal-bound CO are sensitive probes of binding sites due to the high density of states of this ligand around the highest occupied and lowest unoccupied orbitals i.e. around the Fermi level, E_F. Such a partial density of states is basically a projection of the electronic states of the united system onto the wave-functions of the separated species; for metal-bound CO onto the orbitals of a free CO molecule. Normally CO binds with the carbon end towards the metal atom, allowing electrons to be transferred from the CO 5σ orbital to the empty adsorbent states ($3d\sigma$) and back-donated from the metal ($3d\pi$) to the antibonding CO 2π orbital. The so-called π-back-bonding, is crucially dependent upon the electronic structure of the adsorbent and is, for surface-bound CO, markedly altered by "poisonous" electronegative or "promoting" electropositive adsorbates. The origin of this influence is a matter of current interest. One model emphasizes short-range or local interactions (<2 interatomic distances) between COπ orbitals and alkali ns orbitals [1] whereas another, "non-local" model [2] attempts to correlate work-function shifts and vibrational frequencies; $\dfrac{\Delta \nu_{CO\ chem}}{\Delta \phi} > 0$ [3].
The latter correlation refers to the macroscopic work-function as defined, for example, by the threshold energy of photoionization. Literature also defines a "local work-function" or electrostatic potential as will be discussed below.

The electronic structure of metalloporphyrins is similar to that of a metal surface in that the metal atom is supported by a buffering supply of delocalized electrons and the 3d-states are comparatively localized. Iron atoms chelated by porphyrins form the active sites of hemeproteins. The vibrational frequencies – ν_{CO} and ν_{FeC} – of liganded CO, and the ferric/ferrous reduction potential, E_0', respond to the electronegativity of peripheral substituents of the heme (ironporphyrin) plane as well as to perturbations via the protein-porphyrin interface [4].

In this communication I will compare the electronic structures of metal surfaces with that of metalloporphyrins and discuss the possibility of isolate non-local electronic perturbations for CO terminally bound to metal surfaces from the vibrational spectra of carbonyl-hemes.

2. Electronic Structure

A. Metal Surfaces

The outer-shell electrons of transition metals are separated into two bands. The filling of the narrow d-band corresponds to the occupation of the d-shell, whereas a broad sp-band is produced by the valence electrons of the free atom. d-states are comparatively non-dispersive, the electrons being rather localized, whereas sp-states are more free-electron like. In real space this means that whereas d-electrons to a large proportion reside around a certain atomic core or within a certain Wigner-Seitz cell sp-electrons penetrate the entire bulk [5]. Hence a comprehensive theoretical model of a metal surface has to account for delocalized electron states as well as for the discrete atomic character of the adsorbent.

The reactivity of metal surfaces relates to several properties. One is the metal density of states at the Fermi level $N(E_F)$, another the symmetry of available adsorption sites, and a third the chemical potential. For most metal surfaces, in particular for the coinage metals, the chemical activity is markedly influenced by defects. Adatoms introduce radical-like states pinned at E_F, for Cu of σ ($4s$) symmetry and for Fe of σ and π ($3d$) symmetries [6]. On the other hand adsorbed alkali atoms, well known promotors of CO dissociation, exert their influence in a different manner. Calculations have lent support both to models based on a non-local influence [7, 8] and to models where the influence on adsorbed CO molecules is restricted to nearest neighbouring sites around alkali atoms [1, 9, 10, 11].

An antisymmetric combination of ns orbitals on two alkali atoms can couple to the CO 2π orbital. Formally this perturbation lifts the π degeneracy since the combined system has C_{2v} symmetry. The energy denominator should be small for such a perturbation since both the alkali ns and the CO 2π orbitals are located close to E_F. If the spacial overlap between these orbitals is significant i.e. if the species occupy adjacent sites on the adsorbent a considerably enhanced occupation of the 2π level should follow [1]. In another "local" approach Tomanek and Bennemann show that a K atom, as a substitutional impurity in the surface layer of Ni(111), will decrease the activation barrier for CO dissociation if the molecule is bound parallel to the surface [9]. They represent the clean surface by a closely packed cluster of four atoms and the substituted surface by Ni_3K. In a competing approach Ray and Andersson simulate a non-local effect of K adsorbed onto Pt(111) by a varying Pt valence state ionization potential [7]. They predict an altered most favourable adsorption site and weakened CO force constants for all sites as results of a lowered ionization potential, i.e. an enhanced K coverage. Their predictions are in accordance with experimental results for Pt(111)/K/CO [2].

Two more calculations on the character of influence by "poisonous", electronegative adsorbates give apparently contradictory results. Feibelman and Hamann found an effect on the Fermi level density of states two adsorbent atoms away from a sulphur atom adsorbed onto Rh(001) [8]. They used a self-consistent linearized-augmented-plane-wave scheme whereas in the second calculation Joyner et al. use a Green function method to investigate the perturbations induced by carbon atoms adsorbed onto Ni(100) [10]. Both calculations agree upon the short-range character of the net induced density of states by electronegative adsorbates but Joyner et al. found no effect upon $N(E_F)$ at second nearest neighbouring sites. A recent calculation for Rh(001)/Li indicates that electropositive promotors may cause a non-local change of the total electron density as well as of $N(E_F)$ [12].

Calculations quoted above represent the surface either by a cluster of atoms or by a slab of atoms with translational symmetry. The jellium and effective medium approaches emphasize instead the delocalized character of the electron gas. Nørskov et al. conclude that the poisonous or promoting effects of adsorbates should be related to the sign of the electrostatic potential around the adsorbed atom, i.e. a short-distance effect [11]. This is somehow similar to the situation when an external electric field "pulls" the CO density of states down, thus enhancing the filling of the anti-

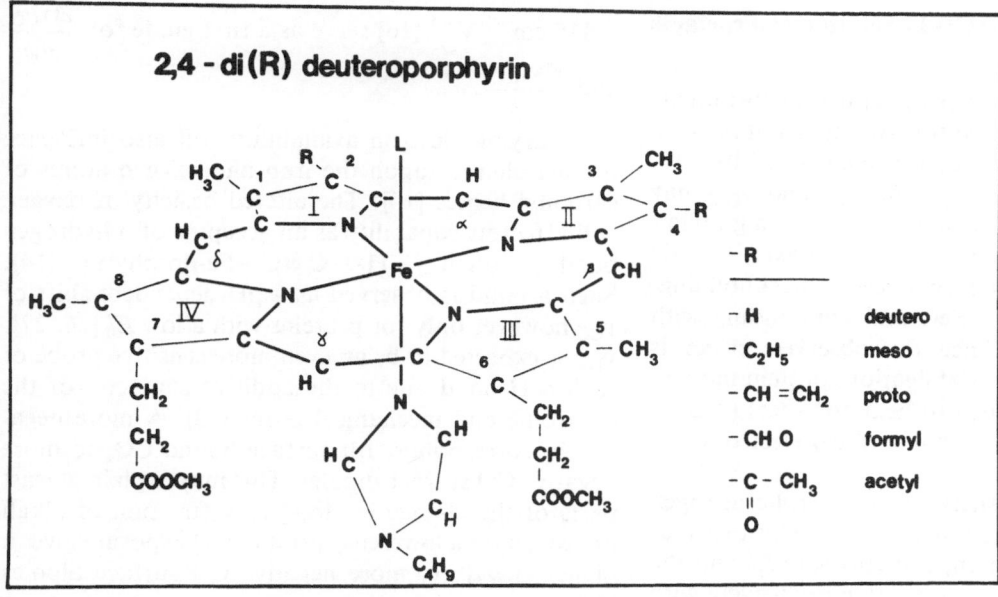

Fig. 1. 2,4-Substituted deutero heme (ironporphyrin) dimethyl ester. *L* marks the position of liganded CO

bonding 2π level [13, 14]. The initial state energies E-E_F of the $\mathrm{Xe}\,5p_{1/2}$ and $\mathrm{Xe}\,5p_{3/2}$ orbitals monitor the "local work-function" or electrostatic potential [15]. Hence physisorbed Xenon atoms ought to be informative probes of promoted metal catalysts.

Experimental studies of surface-bound CO can hardly separate "local" and "non-local" contributions since even at low alkali coverages a considerable number of possible CO adsorption sites are disqualified as nearest neighbours to alkali atoms and thus close enough for local CO-alkali interactions. Hence interpretations of experimental data are controversial and gain support for both models [16 plus references therein].

B. Metalloporphyrins

Porphyrins (fig. 1) are identified by their electronic transitions. As with other conjugated hydrocarbons strong absorption bands are found in the ultraviolet. These bands stem from excitations between delocalized "π" orbitals fairly well modelled by free electron states [17]. The electron structure of the metal atom is usually discussed in terms of d orbitals in an effective ligand field. Depending upon the state of oxidation, $(2+n)$, and the strength of the field, iron is assumed to be a "term" within a Fe $3d^{(6-n)}$ configuration [18]. It is convenient to discuss spin-states and electron spin resonance spectra in such a model. Ferrous iron forms a low spin $t_{2g}^6 e_g^0 (^1A_{1g})$ configuration in a strong ligand

field of octahedral symmetry [18]. This is the stable configuration in hexacoordinated carbonyl-hemes.

The catalytic specificity of heme-proteins is partly determined by a delicate adjustment of the energy required for the partitioning of electrons from the metal atom, i.e. in the initial state the energy of the highest occupied molecular orbital. The protein moieties exert their influence, to some extent at least, by imparting slightly different Fe(III)/Fe(II) reduction potentials (E_0') to the hemeproteins. Some hemeproteins may be reconstituted with an artificial heme i.e. iron-porphyrins with different substituents R (fig. 1).

Barlow et al. [19] and later Smith et al. [4] found a linear correlation between E_0' (expressed as E_{m7} i.e. E_0' at pH 7 and 50% reduction) and the CO stretching frequency, ν_{CO}, of carbonyl-hemes. A lowered E_0' was accompanied by a lowered ν_{CO}, the slope being around $50\ \mathrm{cm}^{-1}\mathrm{V}^{-1}$. This correlation holds not only if the iron atom is perturbed by peripheral substituents on the 2,4 positions but also if the interactions originate from different proteins. 2,4-Substituents exert their influence by long-range electronic perturbations via delocalized states on the porphyrin [16]. The protein influence is less well understood. Stellwagen found a linear correlation between the protein-porphyrin interfacial area and E_0' [20]. This observation indicates that non-covalent interactions between the protein moieties and the porphyrin electronic states may express themselves as a drift of E_0'. Other models emphasize steric or electrostatic influence by the ligand trans to the active binding site L (fig. 1) [21].

3. Model adsorbents for CO bound to metal surfaces

The electronegativity of peripheral 2,4 substituents of the heme govern the electron availability at the iron atom such that a lowered E_0' is accompanied by a reduced ν_{CO} and a raised ν_{FeC} [16]. ν_{CO} and ν_{FeC} may correlate for two reasons; either because of a gradually enhanced bending of the $Fe-C-O$ axis, a steric effect, or because of an enhanced π-backbonding, bonding with respect to FeC and antibonding with respect to CO [22]. Since the observed effect is additive irrespective of axial ligation, protein moiety, and solvent it is concluded to be a non-local (>2 interatomic distances in a metal surface) electronic effect.

Furthermore, the basicity of the pyrrolic nitrogen atoms correlate with these changes [23]. This observation disqualifies any attempt to correlate E_0' and the electronic structure of the iron atom exclusively with the axial ligand strength, simply because there are no axial ligands when the iron atom is replaced by the two hydrogen atoms. Instead this observation favours a model based on a varying electronic structure at the pyrrolic nitrogens, at least. In such a model ligand field variations between hemes in different proteins seem to be of secondary importance compared to charge delocalization [4]. Obviously $3d$ orbitals will respond to a change of the crystal field, but the marked influence on the vibrational frequencies of liganded CO stems from delocalized $4p$ electrons [4]. This electron delocalization is paralleled by a non-local influence of adsorbed alkali atoms on the binding sites of metal surfaces. The basis for this comparison is charge-accommodation, in porphyrins by free-electron like "π" orbitals and in metal surfaces by sp electrons. The presence of delocalized electronic states in metalloporphyrins make them akin to metal surfaces rather than to, for example, metal carbonyls or zeolites. In the absence of delocalized "π" electrons and with a less rigid metal embedding no such analogies to metal surfaces are likely to apply to non-porphyrin metalloproteins. The strength and symmetry of the ligand field determine the cupric/cuprous reduction potential of blue copper proteins [24].

The reduction potential, E_0', or the closely related electrochemical potential equals the ionization potential of the reduced specie plus solvent effects [25]. On the other hand the ionization potential of a metal surface i.e. the work function, ϕ, equals the chemical potential plus the surface barrier [25]. Hence the empirical ratios, $\dfrac{\Delta \nu_{CO}}{\Delta E_0'} = 65\ \mathrm{cm}^{-1}\ \mathrm{V}^{-1}$ and $\dfrac{\Delta \nu_{FeC}}{\Delta E_0'} =$

$-119\ \mathrm{cm}^{-1}\ \mathrm{V}^{-1}$ [16] serve as a first guide for $\dfrac{\Delta \nu_{CO}}{\Delta \phi}$ and $\dfrac{\Delta \nu_{MeC}}{\Delta \phi}$.

A varying electron availability will also influence the net charges upon the iron and oxygen atoms of carbonyl hemes [14]. The altered basicity of oxygen will affect its capability as an acceptor of a hydrogen bond, protein $-H\cdots O \equiv C - Fe$-porphyrin [14]. Such a bond is observed as a pH-dependent shift of ν_{CO} however only for proteins with a low E_0' [26, 27]. ν_{FeC} is expected to be an even more sensitive probe of such a H-bond due to the additive character of the electronic and mechanical terms [28]. A more negative q_0 corresponds, for surface bound CO, to more negative CO surface dipoles. This may explain at least parts of the change in $\Delta \phi_{CO}$ as a function of alkali preexposure; a lowered ϕ prior to CO exposure gives a higher $\Delta \phi_{CO}$ i.e. more negative CO surface dipoles [29].

Independent proofs of a varying electron density at iron, e.g. as a function of 2,4 substituents, may be given by chemical shifts of core level excitation energies. ESCA studies on ferrous carbonyl compounds cannot be performed due to the high rate of CO dissociation. More stable ferric compounds, e.g. μ-oxobishemins, are on the other hand less sensitive to the relatively small perturbations by peripheral substituents. EXAFS edge studies may thus prove to be a more efficient tool. ESCA studies of BF_4 doped polypyrrole have shown that the removal of one electron from each unit of four pyrrol groups is completely smeared out over the entire unit much in the same manner as is expected in the present "electron-gas" model of substituted hemes [30].

Obviously there are differences between metal surfaces and metalloporphyrins as well. The electron gas of a metal contains moles of electrons whereas that of a porphyrin is restricted to approximately 25 electrons. This gives a continuous density of states for a metal but discrete bands for a porphyrin. Electron stoichiometry, which is essential in the redox reactions of proteins, is less relevant on metal surfaces, since the "Fermi sea" is a practically unlimited buffer of electrons. The substitution of an amino-acid in the protein may alter the relative affinities for carbon monoxide and dioxygen between two hemoglobins without much, or any, shift of E_0'. This is perceived as steric influences or direct impacts on the liganded molecule from the surrounding proteins. Moreover, metal surfaces provide binding sites of different symmetries, which can be highly selective in certain reactions. Nevertheless, mainly because of the response of the vibrational bands of liganded CO to the electron availability of hemeproteins I regard hemes as relevant models for terminal bonding on metal surfaces. I conclude that the same pattern should apply to other multiple bonded ligands (O_2, N_2) when the metal atom is supported by a buffer of electrons.

Acknowledgements

Parts of this work were done in cooperation with Dr M. L. Smith and Prof. K. G. Paul, Dept. of Physiological Chemistry, University of Umeå.

References

1. Brodén G, Gafner G, Bonzel HP (1979) Surf Sci 84:295
2. Garfunkel EL, Crowell JE, Somorjai GA (1982) J Phys Chem 86:310
3. Nieuwenhuys BE (1981) Surf Sci 105:505
4. Smith ML, Paul J, Ohlsson PI, Paul KG (1984) Biochemistry 23:6776
5. See e.g. Ashcroft NW, Mermin ND (1976) Solid State Physics. Holt, Rinehart and Winston, New York
6. Paul J, Rosén A (1983) Surf Sci 127:L93
7. Ray NK, Anderson AB (1983) Surf Sci 125:803
8. Feibelman PJ, Hamann DR (1984) Phys Rev Lett 52:61
9. Tomanek D, Bennemann KH (1983) Surf Sci 127:L111
10. Joyner RW, Pendry JB, Saldin DK, Tennison SR (1984) Surf Sci 138:84
11. Nørskov JK, Holloway S, Lang ND (1984) Surf Sci 137:65
12. Feibelman PJ, Hamann DR (1985) Surf Sci 149:48
13. Holloway S, Nørskov JK (1984) J Electroanal Chem 161:193
14. Paul J, Rosén A (1984) Chem Phys Lett 105:197
15. Wandelt K (manuscript)
16. Paul J (1984) Thesis, Chalmers Univ. of Tech., Göteborg
17. Goutermann M (1961) J Mol Spectr 6:138
18. Lever ABP, Gray HB (eds) (1983) Iron Porphyrins, Part 1 and 2. Addison-Wesley, Reading
19. Barlow CH, Ohlsson PI, Paul KG (1976) Biochemistry 15:2225
20. Stellwagen E (1978) Nature 275:73
21. Moore GR, Williams RJP (1977) FEBS Lett 79:229
22. Blyholder G (1964) J Phys Chem 68:2772
23. Falk JE (1964) Porphyrins and Metalloporphyrins. Elsevier, Amsterdam
24. Gray HB, Malmström BG (1983) Comm Inorg Chem 2:203
25. Gerischer H, Ekardt W (1983) Appl Phys Lett 43:393, plus references therein
26. Smith ML, Ohlsson PI, Paul KG (1983) FEBS Lett 163:303
27. Satterlee JD, Erman JE (1984) J Amer Chem Soc 106:1139
28. Paul J, Paul KG (unpublished)
29. Lindgren SÅ, Paul J, Walldén L (1982) Surf Sci 117:426
30. Salaneck WR, Erlandsson R, Prejza J, Lundström I, Inganäs O (1983) Synth Metals 5:125

Received June 26, 1984;
accepted October 15, 1984

Author's address:

J. Paul
Exxon Research and Engineering Company
Clinton Township
Route 22 East
Annandale NJ 08801 (U.S.A.)

Progress in Colloid & Polymer Science   Progr Colloid & Polymer Sci 70:62 – 66 (1985)

Bilayer formation at adsorption of proteins from aqueous solutions on metal surfaces

T. Arnebrant, B. Ivarsson*, K. Larsson, I. Lundström*, and T. Nylander

Department of Food Technology, University of Lund, Lund (Sweden)
*Laboratory of Applied Physics, Linköping Institute of Technology, Linköping (Sweden)

Abstract: Studies of adsorption of β-lactoglobulin and ovalbumin on plasma cleaned chromium surfaces by ellipsometry show that a thick highly hydrated layer is obtained, which can be partly removed by aqueous buffer rinsing. Changes in the electrode potential during the adsorption and rinsing process were also followed. The results indicate that the protein adsorbs into a bilayer, with the bottom layer unfolded and attached by strong polar bonds to the surface, so that it is not removed by rinsing. On top of that layer, protein molecules are attached by hydrophobic interaction and/or ionic forces. The upper layer results in large electrical changes. For comparison the same measurements were done on chromium surfaces, which were made hydrophobic, and these results are in agreement with the generally accepted model of monolayer adsorption.

Key words: Protein adsorption, metal-water interface, protein bilayer, ellipsometry, potential measurements

Introduction

Contacts between aqueous protein solutions and metal surfaces take place in many applications of technical or medical interest. Examples are fouling on heat exchangers in the food industry, the behaviour of foreign materials like medical implants *in vivo* and surface orientated diagnostic methods.

There are also fundamental reasons for the study of protein molecules on surfaces. They are complicated molecules with hydrophilic as well as hydrophobic regions, which can undergo conformational changes upon adsorption. From extensive studies of protein adsorption on hydrophobic materials [cf. ref. 1] it has been concluded that the adsorption isotherms develop plateau values, which correspond to monolayers of varying packing density. In the case of extreme protein concentration, "interfacial coagulation" [2] has been observed. Studies of simple amphiphiles and their adsorption on metal oxides, on the other hand, has shown that the plateau values of the adsorption isotherms correspond to a bilayer of the amphiphile molecules [3]. There is less information on the adsorption behaviour of proteins. The present paper is based on ellipsometry studies and potential measurements, and we suggest that a bilayer of protein is formed in the case of clean metal surfaces. The first

layer is irreversibly adsorbed on the metal surface, whereas the second layer desorbs upon rinsing.

Experimental

Desalted ovalbumin prepared according to [4] and β-lactoglobulin (L6879, lot no 111F-8025), Sigma Chemical Co, desalted by dialysis against distilled water and freeze-dried, were used. The proteins were dissolved in phosphate buffered saline (PBS) (pH = 7.0, 0.01 M phosphate, 0.15 M NaCl, ionic strength = 0.21 M) [5] and the solutions were used within one day.

Glass slides with vacuum deposited chromium on a rectangular area of 0.5 cm^2 were used. The metal film thickness was about 3000 Å. The chromium plated slide was made hydrophobic by repeated washing with a detergent solution, double distilled water, and finally chloroform. This treatment was found to give a hydrophobic surface with a critical surface tension of about 45 mN/m [6]. The hydrophilic surface was initially prepared in the same way as the hydrophobic. It was then plasma cleaned for 4 minutes in air (0.3 Torr), using a radio frequency glow discharge apparatus (Harrick PDC 3XG). The surface obtained was hydrophilic as obvious from water wettability.

The thickness and the refractive index of the adsorbed protein layer were determined by ellipsometry [7]. The adsorbed mass, Γ, was calculated according to Cuypers et al. [8]. The values used for the molar weight/molar refractivity were 4.10 and 4.17 g/ml and for the partial specific volume 0.751 and 0.750 ml/g for β-lactoglobulin and ovalbumin, respectively. The change in electrode potential, ΔE, was simultaneously measured relative to a standard calomel electrode (SCE). A more detailed description of the methods used in this work is given in [5].

The chromium slide was placed into a sample cell (cuvette) containing 12.6 ml of PBS. When the potential was stable (at the rest potential) a rinsing of pure buffer was done to test the cleanliness of the rinsing system. If the electrode then continued to show ellipsometric and electric stability, 1.4 ml of the protein solution (10 mg/ml) was injected, giving a final protein concentration of 1 mg/ml. After 60 minutes, rinsing was performed through a continuous flow of 200 ml of PBS through the sample cell. A second addition of protein was done when the measured parameters were stable in the new fresh buffer.

All experiments were performed at 25 °C.

Results

The mean values of three representative experiments were used. The maximum deviations were within about 10% both for the adsorbed amount and for the potential change. The adsorbed amounts calculated according to de Feijter et al. [9], using a refractive index increment of 0.183 ml/g, were generally 10% lower than those according to Cuypers used in this work.

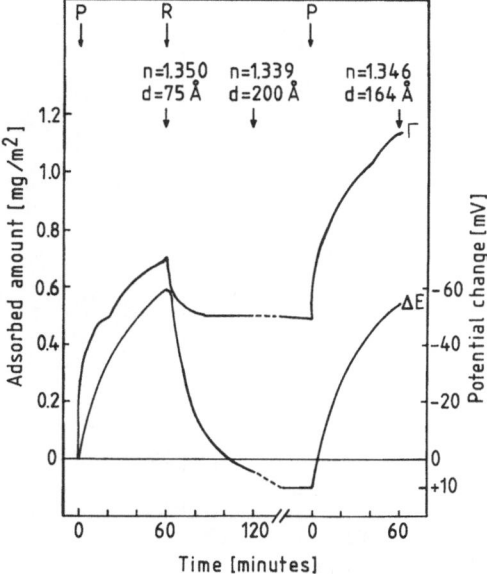

Fig. 2. Adsorbed amount (Γ) and potential change (ΔE) for β-lactoglobulin on a hydrophilic chromium surface. The conditions are the same as in figure 1

The mean values ± max. deviation of the open cell rest potentials were -19 ± 135 and $+45 \pm 20$ mV (vs. SCE) for the hydrophilic and hydrophobic electrodes, respectively. The electrode potential has in several studies been shown to influence protein adsorption (cf. ref. 5). Therefore the range of the rest potential of the hydrophilic surfaces in the present study, from -154 to $+116$ mV vs. SCE could be large enough to influence the adsorption of the two proteins used. These variations in rest potential, however, had no correlation with the spreading of the ellipsometric date and the change in potential.

The adsorbed amounts of β-lactoglobulin, with plateau values of the thickness and of the refractive index, on hydrophobic and hydrophilic chromium surfaces, respectively, are shown in figures 1 and 2. The potential changes are also shown in these figures. The corresponding curves were recorded for ovalbumin and are shown in figures 3 and 4.

The most striking feature is the large decrease in potential change upon rinsing although there is only a minor change in the adsorbed amount of protein. Furthermore the pronounced difference in the nature of the adsorbed protein film is remarkable in the case of hydrophilic and hydrophobic surfaces according to the thickness and refractive index, especially for ovalbumin. These results shown that there are fundamental differences between protein adsorption and de-

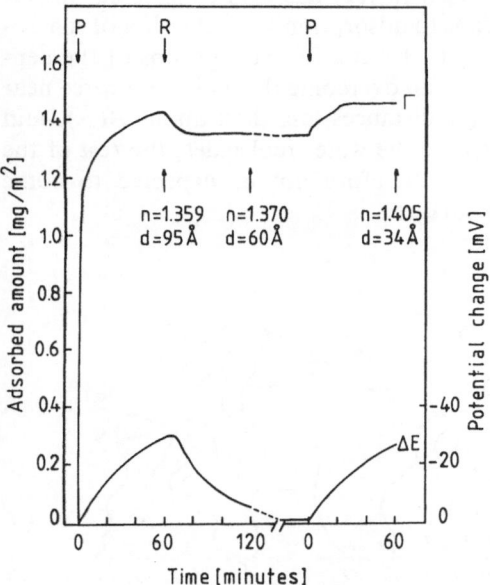

Fig. 1. Adsorbed amount (Γ) and potential change (ΔE) for β-lactoglobulin on a chromium surface which was made hydrophobic. Plateau values of refractive indices (n) and thicknesses (d) are inserted. P and R indicate addition of protein and rinse respectively. In order to approach equilibrium 20 to 50 minutes were allowed (dotted line) prior to the second protein addition

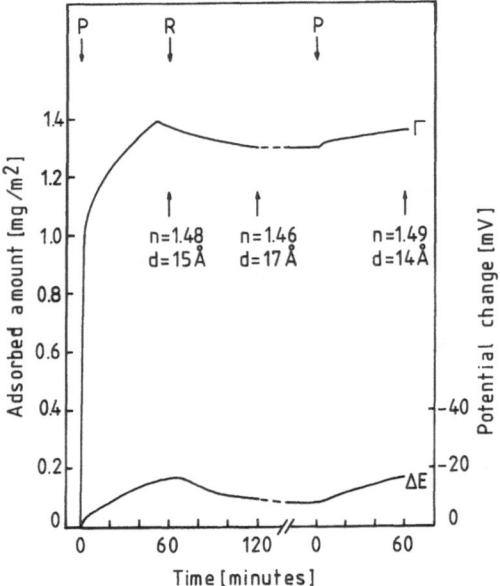

Fig. 3. Adsorbed amount (Γ) and potential change (ΔE) for ovalbumin on a chromium surface which was made hydrophobic. The conditions are the same as in figure 1

sorption on clean metal surfaces compared to the corresponding processes on hydrophobic surfaces, and a model explaining these results will be presented below.

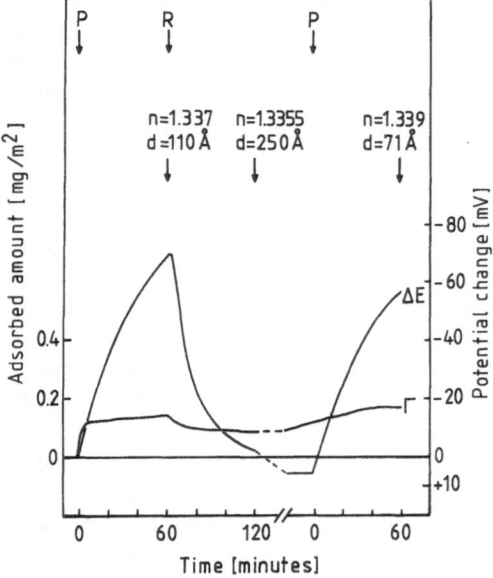

Fig. 4. Adsorbed amount (Γ) and potential change (ΔE) for ovalbumin on a hydrophilic chromium surface. The conditions are the same as in Fig. 1. It should be pointed out that the refractive index is given with high accuracy as it is so close to that of the buffer solution

Discussion

The adsorption and desorption of the globular proteins, which were investigated, indicate a bilayer mechanism of protein adsorption on clean metal surfaces. The different steps in this proposed model will first be described, and then the experimental results will be considered in relation to this model.

There are always some polar amino acid side chains of a globular protein, which can interact strongly with a metal surface, even if both the protein and the metal surface in many applications are negatively charged. Such binding should be expected to result in unfolding of the protein and therefore also in an irreversible adsorption. Another consequence must be the exposure of loops with hydrophobic character into the aqueous solution. This in turn means that the adsorption of a second protein layer can reduce the free surface energy. A schematic illustration of this model of bilayer adsorption is indicated in figure 5.

It is interesting to relate the highly hydrated protein layer formed on the clean metal surface with recent findings on hydration forces [cf. refs. 10, 11]. It has thus been found that hydrophilic surfaces and colloidal aggregates with hydrophilic surface groups exhibit an exponential repulsive force at short distances which is much stronger than the electrostatic repulsion, and Parseigan [12] has also considered this effect as relevant to adsorption and adhesion of macromolecules. Apart from a few side groups of the peptide chain that can overcome the hydration force near the surface (at distances less than about 20 Å), and thus can displace the water molecules, the rest of the protein should therefore not be expected to come close to the surface.

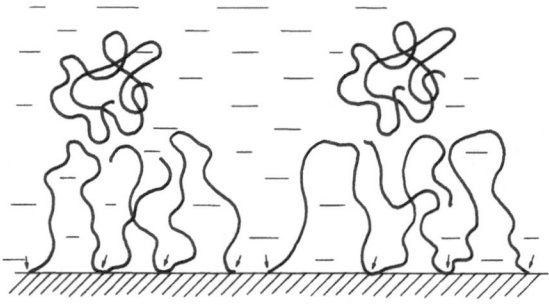

Fig. 5. Schematic illustration of the proposed protein bilayer formation at a hydrophilic surface after adsorption from aqueous buffer solution. The protein structure is suggested to be closer to the native state in the upper layer than in the irreversibly bound bottom layer. The arrows indicate bonds between polar groups of the protein molecule and the surface

In the case of a negatively charged metal surface, for example, it should be expected that positive groups like the N-terminal and lysine residues bind with direct contacts to the surface. Such charge compensation may explain why the electrical effects of this bottom monolayer appear relatively weak, compared with the upper layer. It is also rather obvious that ionic interaction beside hydrophobic interaction can link groups in each of the protein layers. Thus in the case of a negative surface charge and protein molecules with a negative net charge, it is obvious that the upper layer will tend to orient its positive residues downwards and the negatively charged side-chains outwards from the surface. Due to the bulkiness of the protein molecule, a stronger dipolar effect might therefore be expected for the upper layer. In this connection it should be mentioned that the isoelectric points are 5.2 and 4.6 for β-lactoglobulin and ovalbumin, respectively.

From the successive steps of protein adsorption described above it should be expected that the whole layer is highly hydrated, and the refractive index values close to that of water, as well as the high thickness values are consistent with this model. The binding forces between the hydrated protein monolayers can hardly be comparable in strength to the binding at the metal surface, and it is therefore reasonable to expect that rinsing in aqueous buffer results in desorption of the upper layer, and the reduction in adsorbed mass as well as the strong reduction in the potential change appear to be qualitatively consistent with the model.

Thus we regard the hydrated protein double-layer as a two-dimensional gel-phase, with molecular linkages formed by hydrophobic interaction and/or ionic bonds. The rinsing experiments show that the upper protein layer is loosely attached, which indicates that this outer layer has a structure closer to the native state. The reduction in adsorbed amount and the electrical changes upon rinsing are in good agreement with such a structure.

Simple kinetic arguments support our view about the potential changes being due to mainly a second protein layer. They are discussed in another contribution [13] in this issue.

Based on adsorption studies using radioactive labelled proteins and other techniques, models of the formation of protein films and their structure have been proposed earlier [cf. 14 – 20]. Most relevant to this study is the work by Brash et al. [18, 19], where indication of a second layer of negatively charged fibrinogen on a positively charged polymer surface is found. Under these conditions it seems probable that

the ionic forces responsible for the primary adsorption will change the conformation of the molecules and expose hydrophobic sites, which will induce adsorption of a second layer. Behaviour similar to that described above in the case of metal surfaces should be expected when ionic forces are involved in the adsorption process.

The combined measurements of adsorbed amount and electrical effects in the present work indicate that monolayer models are not adequate to explain our observations. In analogy with the behaviour of the lipid type of amphiphiles, which adsorb as bilayers on metal oxides [3], the bilayer structure is a natural model to consider also in the case of amphiphilic proteins.

In the case of proteins exposed to a hydrophobic metal surface, the behaviour is quite different, and in good agreement with earlier studies of protein adsorption to hydrophobic materials [1]. Thus the values of the thickness and of the refractive index are consistent with the formation of a monolayer.

What are the consequences of a second protein layer at a hydrophilic metal surface? Since this layer is weakly associated, there will be a rapid exchange of proteins between the solution and the surface. If the loosely bound protein molecules change their conformation irreversibly upon association with the strongly bound bottom layer molecules, a number of protein molecules with modified conformation will enter the solution. If this is the case when a metal is implanted *in vivo,* immunological effects can occur. We therefore conclude that the observed physical phenomenon may be of significance for the understanding of biocompatibility.

Acknowledgement

Financial support was obtained from the Swedish Board for Technical Development. Thanks to Mrs Gunnel Lundh for preparing the figures.

References

1. Norde W, Lyklema J (1979) J Colloid Interface Sci 71:350
2. MacRitchie F (1972) J Colloid Interface Sci 38:484
3. Scamehorn JF, Schechter RS, Wade WH (1982) J Colloid Interface Sci 85:463
4. Hegg P-O (1979) Biochim Biophys Acta 579:73
5. Ivarsson BA, Hegg P-O, Lundström I, Jönsson U. accepted for publication in Colloids Surf
6. Arnebrant T, Nylander T, Cuypers PA, Hegg P-O, Larsson K (1983) In: Mittal KL, Lindman B (eds) Surfactants in Solution, Vol. 2. Plenum Press, New York, p 1291

7. McCrackin FL (1969) A Fortran Program for Analysis of Ellipsometer Measurements. National Bureau of Standards Technical Note 479, Washington
8. Cuypers PA, Corsel JW, Janssen MP, Kop JMM, Hermens WTh, Hemker HC (1983) J Biol Chem 258:2426
9. De Feijter JA, Bejamins J, Veer FA (1978) Biopolymers 17:1759
10. Le Neveu D, Rand RP, Parseigan VA, Gingell D (1977) Biophys J 18:209
11. Israelachvili J, Pashley RM (1983) Nature 306:249
12. Parseigan VA (1982) Adv Colloid Interface Sci 16:49
13. Lundström I (1985) Progr Colloid Polymer Sci 70:76
14. Dillman Jr WJ, Miller IF (1973) J Colloid Interface Sci 44:221
15. Chan BMC, Brash JL (1981) J Colloid Interface Sci 82:217
16. Bornzin GA, Miller IF (1982) J Colloid Interface Sci 86:539
17. LeCompte M-F, Rubenstein I, Miller IR (1983) J Colloid Interface Sci 91:12
18. Schmitt A, Varoqui R, Uniyal S, Brash JL, Pusineri C (1983) J Colloid Interface Sci 92:25
19. Brash JL, Uniyal S, Pusineri C, Schmitt A (1983) J Colloid Interface Sci 95:28
20. De Baillou N, Dejardin P, Schmitt A, Brash JL (1984) J Colloid Interface Sci 100:167

Received June 26, 1984;
accepted October 15, 1984

Authors' address:

Thomas Arnebrant
Department of Food Technology
University of Lund
Box 124
S-22100 Lund (Sweden)

Progress in Colloid & Polymer Science   Progr Colloid & Polymer Sci 70:67 – 75 (1985)

Fourier transform infrared reflection absorption spectroscopy (FT-IRAS) of some biologically important molecules adsorbed on metal surfaces

B. Liedberg, B. Ivarsson, I. Lundström, and W. R. Salaneck*

Laboratory of Applied Physics and *Laboratory of Surface Physics and Chemistry,
Linköping Institute of Technology, Linköping (Sweden)

Abstract: The technique of Fourier transform infrared reflection absorption spectroscopy (FT-IRAS) has been used to characterize thin films of proteins and amino acids formed on metal surfaces. The films were spontaneously adsorbed from solution onto each metal. All infrared spectra were taken on dried films. The infrared spectra were obtained by reflecting the radiation from the film co-vered surface at a high angle of incidence ($\sim 75°$) and with the radiation polarized parallel to the plane of incidence.

The IRAS spectra of fibrinogen adsorbed on gold, copper, aluminium and titanium all showed a large blue shift of the Amide I frequency ca. 20 cm^{-1}, relative to the same band in the aqueous and solid phases. Similar shifts were also observed for lysozyme and pepsin adsorbed on these metal surfaces. An even larger shift ca. 32 cm^{-1} was observed for β-lactoglobulin adsorbed on gold and titanium. These distinct shifts indicate that conformational change of the proteins occurs upon adsorption on metal surfaces, however, other alternative explanations for these blue shifts are also discussed.

In addition, model experiments of the smallest amino acid glycine on gold indicate that the molecule, in its zwitterionic form, adsorbs with a preferred orientation on the surface such that the NH_3^+ group is close to the gold surface.

Key words: FT-IR, IRAS, proteins, biocompatibility, amino acids

Introduction

Infrared spectroscopy has been extensively used in studies concerned with the interaction between biological macromolecules such as proteins and foreign artificial materials [1 – 4]. A variety of protein-polymer systems has been investigated by using the surface sensitive ATR-technique (attenuated total reflection). In these studies, the protein films were formed on bare or polymer coated ATR-crystals by adsorption from protein mixtures in aqueous solution or form whole blood. It has been shown by various methods that the nature of the protein films, initially formed on foreign surfaces in contact with blood, plays an important role for surface related phenomena such as blood clotting, cell adhesion and thrombus formation [5, 6]. It has also been shown that the plasma protein fibrinogen is the major component adsorbed, at least initially, when a foreign artificial material is brought into contact with whole blood [4, 7].

The limited amount of spectroscopical data available in the literature concerning the interaction be-tween proteins and metals has encouraged us to study fibrinogen and some other proteins adsorbed on metal and metal oxide surfaces. Model metals, such as copper, gold, aluminium and titanium have been investigated. The latter two metals have both been found to be acceptable as implant materials in the human body [8, 9].

A detailed interpretation of protein spectra in general, tends to be very complicated, mainly because of the large variety of side chain bands. These side chains are often terminated by surface active groups such as carboxylate, amino and aromatic groups. In order to obtain a detailed picture of the interaction between these groups and metals, we have begun a study of amino acids adsorbed on metal surfaces. We present some exploratory experiments of glycine adsorbed on gold.

The ATR-technique does not apply to studies of adsorbed molecules on well defined bulk metal surfaces. This is due to the fundamental electrodynamic properties of metals, which do not allow the electric field to penetrate the metal to any significant extent at infrared wavelengths. An external reflection technique is apparently necessary for obtaining IR-spectra of adsorbed molecules on metals. Infrared reflection absorption spectroscopy has proven to be very useful in many areas of surface chemistry. On great advantage of this technique is that it can produce direct information of the geometrical arrangement of the adsorbed molecules relative to the metal surface. The ATR-technique and reflection absorption spectroscopy are in a sense complementary, where the former technique is superior for studies of adsorbed molecules on highly infrared transparent materials, such as polymers, while the latter is more suitable for applications on metals. One advantage of the ATR-technique over reflection absorption spectroscopy is that it can be more easily applied to *in situ* studies in an aqueous environment [3, 4].

The technique of infrared reflection absorption spectroscopy (IRAS) was first described both theoretically and experimentally in the late fifties by Francis and Ellison [10]. They used the technique for structural characterization of Langmuir Blodgett films of metal stearates on highly reflecting metal substrates. It is worth noting that IRAS is sometimes called IRRAS or RAIRS. The technique is briefly described in the following, a detailed theoretical treatment can be found elsewhere [10, 11].

A molecule will generally be able to absorb infrared radiation if the molecular vibration is associated with a simultaneous change in dipole moment. In addition to this fundamental "selection rule", Francis and Ellison [10] predicted that the molecule needed to have a component of its dipole moment change (transition moment) normal to the metal surface in order to be infrared active. This is due to the fact that the perpendicular component of the radiation undergoes an ap-

proximate phase shift of 180° for all angles of incidence upon a direct reflection, i.e., the electric field has a node at the surface for the perpendicular component. Accordingly, no net electric field can interact with the oscillating dipoles at the metal surface and thus no absorption can be observed. The parallel components, however, do not cancel because of the finite phase shift between the incident and the reflected components upon reflection. Therefore, a nonzero electric field exists normal to the metal surface, figure 1a. From the discussion above it follows that only vibrational modes with transition moment normal to the metal surface can be excited by the infrared radiation. This additional requirement for excitation of adsorbed molecules on metal surfaces is normally summarized in the "surface selection rule". Figure 1b illustrates this rule for two differently oriented dipoles.

An approximative expression of the fractional change in reflectivity of the parallel component $\Delta R_{||}$ was given by Francis and Ellison [10]

$$\Delta R_{||} = 1 - \frac{R_1}{R_0} = \frac{k \tan \theta \sin \theta}{N^3} \, \varepsilon C d, \qquad (1)$$

where $k = 4/^{10}\log e$. R_1 and R_0 are the reflectivity with and without the film present. θ is the angle of incidence. C and ε are the concentration and the molar extinction coefficient respectively. N and d are the refractive index and the film thickness respectively. The above expression holds for highly reflecting metals such as gold, copper and silver and for angles of incidence between 0° and 80° [10].

Since the early paper of Francis and Ellison in 1959, IRAS has found several applications to surface related phenomena. This includes studies of corrosion-in-

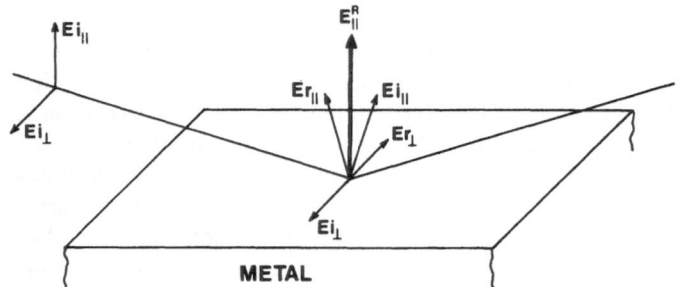

Fig. 1a. The electric field at the metal surface for a high angle of incidence and for both parallel and perpendicular polarization of the infrared radiation. The resulting electric field normal to the metal surface is indicated by $E_{||}^R$. The electric field in the plane of the metal surface is almost zero, since $E_{i\perp} = -E_{r\perp}$

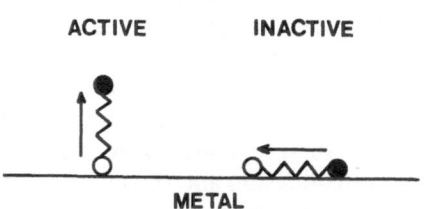

Fig. 1b. Two hypothetical dipoles adsorbed in different orientations on the metal surface. The "surface selection rule" allows only the vibration with dipole moment change normal to the metal surface to be infrared active. The arrows indicate the direction of the transition moment

hibitors on copper [12], structural characterization of fatty acids on copper and iron [13], reactions of carboxylic acids on copper [14, 15] and electrochemical studies of the electrode-solution interface [16, 17]. The dominating application of IRAS, however, has been to small molecules adsorbed onto various metals from the gas phase, under ultra high vacuum conditions (e.g., carbon monoxide on gold and formic acid on nickel) [18–23]. We believe that IRAS can also find further applications in the field of surface chemistry. In the following, we present a study with relevance to biocompatibility [24] and some model experiments of general interest to biochemistry.

Materials and methods

The metals were evaporated onto glass slides ($60 \times 15 \times 1$ mm) to a thickness of about 2000 Å. The metals used in the fibrinogen, lysozyme and pepsin experiments were cleaned in 99.5% ethanol and dried in nitrogen gas before the adsorption process, while a plasma cleaning technique was used in the β-lactoglobulin and glycine experiments. The latter technique seemed to give a more effective cleaning of the metal surfaces. The gold surfaces, which were initially hydrophobic, became hydrophilic after this treatment. The hydrophilic nature of these metal substrates was preserved by storing them in distilled water. In conjunction with the infrared work, we have begun an XPS (X-ray photoelectron spectroscopy) study in order to check the quality of the plasma cleaning technique.

Human serum fibrinogen was purchased from AB Kabi, Stockholm. Lysozyme, pepsin and β-lactoglobulin were obtained from Sigma Chemical Co. Glycine was obtained from AnalaR, BDH Chemicals Ltd. The buffer used in the fibrinogen, lysozyme and pepsin experiments was 8.2 mM Na_2HPO_4/1.8 mM KH_2PO_4/0.15 M NaCl, at pH = 7.4 ± 0.1. All chemicals used therein were of analytical grade.

The metals were symmetrically mounted in a 400 ml beaker (Duran, Jena Glas). The beaker was then filled with freshly prepared buffer and placed in a water bath, at 37 ∓ 1 °C. The metals were allowed to stabilize in the tempered buffer for 20 minutes before the protein was added by pipette into the centre of the volume. It is our experience from electrochemical experiments that it is necessary to allow the metals to stabilize in the phosphate buffer in order to obtain a stable open cell potential (rest potential) and thus an equilibrated surface. Magnetic stirring was performed at 100 rpm. A final fibrinogen solution of 0.4 mg/ml was then allowed to adsorb for 60 ∓ 5 minutes. The final protein concentration was 1 mg/ml in the lysozyme and pepsin experiments. The β-lactoglobulin samples were prepared by adsorption from a 1 mg/ml solution (pH = 7.0 ∓ 0.1) at 20 °C. For a more detailed description of the adsorption process, see [24]. The samples were then copiously rinsed with distilled water and left to dry at room temperature for one week before the IR-analysis.

Glycine was adsorbed onto plasma cleaned gold substrates from a 0.2 mg/ml solution for 5 minutes at 30 °C. A pH near the isoelectric point of glycine (pI = 6.0) was chosen in these experiments in order to keep the zwitterionic

form strongly predominant in the solution. After adsorption the samples were blown dry in nitrogen gas.

The infrared spectra were obtained with a BRUKER IFS-113v Fourier Transform Spectrometer. All experiments were carried out at a constant resolution of 4 cm^{-1}. Figure 2 shows the mirror arrangement used for the IRAS experiments. An additional metal mirror was mounted in front of the sample substrate when multiple reflection spectra were recorded. A ZnSe wire grid polarizer was used to remove the perpendicular part of the infrared radiation.

The FT-IRAS spectra presented in the following are compared with absorbance spectra in the solid and aqueous phases. The solid phase spectra were obtained from either cast films or from powder samples dispersed in a KBr-matrix and pressed to a pellet. Severe experimental difficulties have been associated with infrared spectroscopic studies of proteins in aqueous solution, mainly because of the strong obscuring absorption from water in the Amide I region near 1650 cm^{-1}. These problems can now be overcome by taking advantage of the high signal-to-noise ratio and frequency accuracy of today's FT-IR spectrometers. The procedure used for obtaining solution spectra is shown in figure 3. Figure 3A shows the absorbance spectrum of pure H_2O. This spectrum is digitally subtraced from the absorbance spectrum of 10% solution of β-lactoglobulin, figure 3B. The resulting solution spectrum of β-lactoglobulin after subtraction B – A is shown in figure 3C. By utilizing the large data manipulation capabilities provided by the FT-technique and by using liquid nitrogen cooled detectors, it is possible to obtain high quality spectra as shown in figure 3C within 10 minutes.

Results and discussion

The infrared spectra of proteins and polypeptides comprise essentially four strong vibrational modes associated with the peptide link. These are the Amide A band near 3300 cm^{-1} (ν_{N-H}), the Amide I band near 1650 cm^{-1} ($\nu_{C=O}$), the Amide II band near 1550 cm^{-1} and the Amide III band near 1250 cm^{-1}. The latter two modes are both associated with combined stretch-

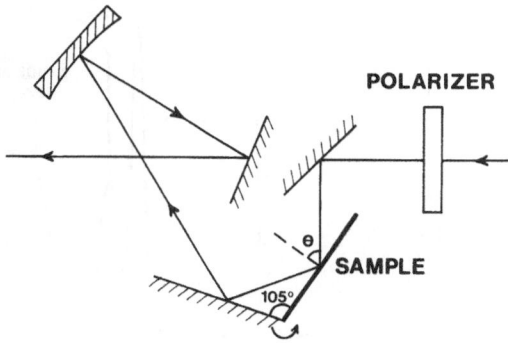

Fig. 2. The optical arrangement used for the IRAS experiments. The polarizer is used to remove the perpendicular part of the infrared radiation

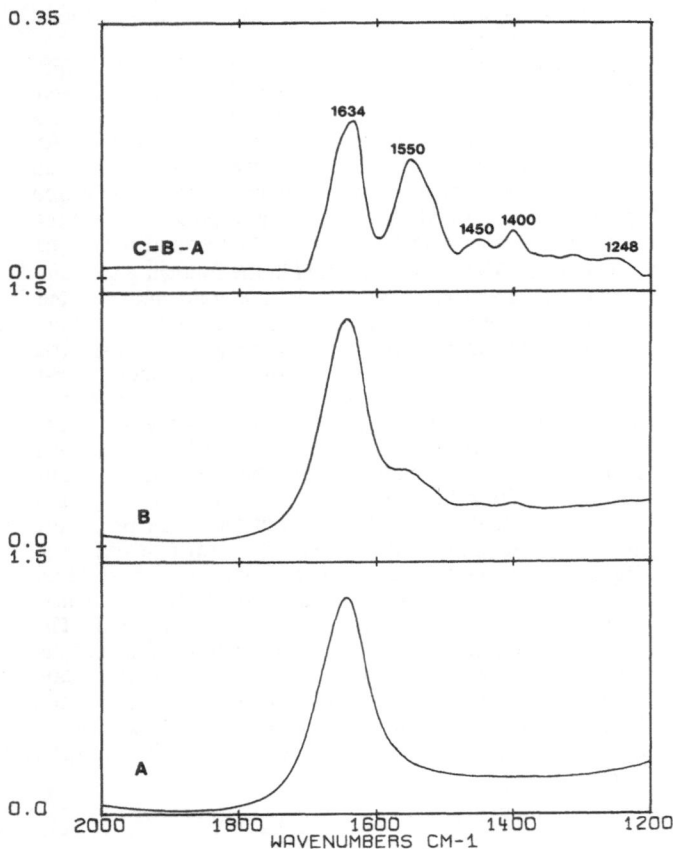

Fig. 3. Aqueous solution spectra in a 6 µm thick CaF$_2$ liquid cell: (A) absorbance spectrum of pure H$_2$O; (B) absorbance spectrum of 10% solution of β-lactoglobulin; (C) the resulting spectrum after subtraction $B - A$

ing and bending vibrations of the C$-$N$-$H bond. The frequency and intensity of the Amide I, II and III bands have been extensively used to determine the chain conformation of proteins and polypeptides [25 $-$ 28]. The side chains give rise to a number of distinct absorption bands, the CO$_2^-$ band near 1400 cm^{-1} and bands near 2900 and 1450 cm^{-1}, which are attributed to stretching and bending modes of CH$_2$ and CH$_3$ groups. For a more detailed interpretation of protein spectra in various physical states, see [28].

Fibrinogen

An overview IRAS spectrum of fibrinogen adsorbed on titanium is shown in figure 4. This figure illustrates that infrared spectra with adequate signal-to-noise ratio can be obtained from monomolecular films on metals by a single reflection technique. The frequencies and assignments of the strongest absorption bands are given in this figure. The strong and broad band in the frequency range from 1200 to 900 cm^{-1} indicates that a large amount of phosphate ions are present on the titanium surface. A similar band pattern is also observed in the aluminium spectrum, while the gold and copper spectra only show minor absorption in this frequency region. These bands are also evident in reference spectra of aluminium and titanium substrates which have been immersed into the phosphate buffer and thereafter copiously rinsed in distilled water. However, this irre-

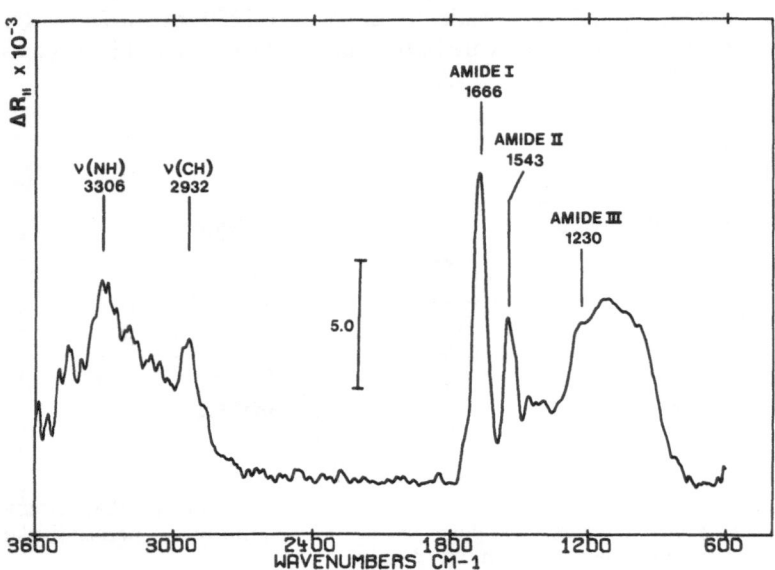

Fig. 4. IRAS spectrum of fibrinogen adsorbed on a titanium surface. One reflection at 75°

versible adsorption of phosphate ions onto the metal oxide surfaces can only partly explain the strong infrared absorption in the 1200 to 900 cm^{-1} region, since all protein spectra are compensated for in this buffer adsorption (i.e. R_0 in equation (1) is actually the reflectivity from a metal substrate which has been immersed into the buffer). Instead, we believe that the phosphate ions are incorporated in the protein films during the adsorption process. The enhanced absorption in the aluminium and titanium spectra may have to do with the fact that these metal oxide surfaces are hydrophilic by nature, while the gold and copper surfaces are more or less hydrophobic. A charge redistribution within a protein adsorbed on the surface is an other possibility, where the phosphate ions are bound to specific sites in the protein.

The Amide I and Amide II bands of fibrinogen are both shifted in frequency upon adsorption on metal surfaces. The Amide I and Amide II frequencies are given in table 1 for fibrinogen in various physical states. The slightly higher Amide II frequency in the solution spectrum compared to the IRAS spectra is probably due to an increased degree of hydrogen-bonding in solution. Similar behaviour of the Amide II band upon dissolution has previously been observed for a large number of proteins [25, 28]. The large blue shift in the Amide I frequency, however, cannot be explained in terms of hydrogen-bonding. Susi et al. [25] and later Koenig et al. [28] showed that the Amide I band of proteins occurs at essentially the same freqency in both the aqueous and solid phases. We conclude, therefore, that the blue shift of the Amide I frequency of spontaneously adsorbed fibrinogen films is induced by the metals. Also evident in table 1 is the fact that *no blue shift* can be observed in the IRAS spectrum of the cast (multilayer) film on gold, even though this film is only a few monolayers thick, indicating that only molecules in close contact with the metal surface are affected. The fact that the Amide I frequency is a very sensitive probe to chain conformation indicates that the fibrinogen molecules

Table 1. The Amide I and Amide II frequencies of fibrinogen in various physical states

Physical state	Amide I (cm^{-1})	Amide II (cm^{-1})
Buffer solution at pH = 7.0	1649	1547
Cast film on Au	1651	1533[a]
Adsorbed film on Au	1670	1543
Adsorbed film on Al(Al$_2$O$_3$)	1672	1543
Adsorbed film on Ti(TiO$_x$)	1666	1543
Adsorbed film on Cu(CuO$_x$)	1669	1543

[a] Diffuse band (uncertain)

undergo a significant degree of structural change upon adsorption onto metals. These structural changes can severely affect the intermolecular and intramolecular interactions between neighbouring peptide groups. A weakening in hydrogen-bonding between peptide groups will be reflected in the infrared spectra as a distinct shift of the Amide I band towards higher frequencies. Charge transfer and dipole-dipole coupling, however, cannot be excluded as alternative explanations. Charge transfer, in particular, may occur between the metal and the protein. An extra negative charge into the peptide group will seriously affect the strength of the $C = O$ bond, which has pronounced single bond character in proteins. This extra negative charge will strengthen the bond and the Amide I vibration will appear at a higher frequency.

The Amide I and Amide II frequencies, however, do not permit a distinction to be made between each individual fibrinogen-metal system (see table 1). The intensities of the Amide I and Amide II bands on each metal are also very similar except for copper, which shows a ten-fold magnification of these bands. The surface concentration (μg/cm^2) of adsorbed protein on gold and copper can be estimated by utilizing equation (1) and by using solution values of the molar extinction coefficient and a refractive index of the film $N = 1.5$ (which is a reasonable value for many organic compounds). The obtained surface concentration can easily be converted to an effective thickness by assuming a protein density of 1.3 g/cm^3 for such films. It is important to note that an effective thickness is not equivalent to a geometrical thickness. The effective thicknesses on gold and copper are approximately 25 Å and 200 Å, respectively. The result on copper is not conflicting with the observation made by Gileadi et al. [29], who postulated from electrochemical measurements on whole blood that copper is an initiator of thrombus formation. Thus large amounts of protein are expected to be adsorbed on a copper surface.

Lysozyme and pepsin adsorbed on gold, aluminium and titanium have also been investigated. Similar changes occur in the infrared spectra upon adsorption of these proteins onto metals. The Amide I bands are shifted approximately 20 cm^{-1} towards higher frequencies and the Amide II bands are shifted to lower frequencies due to the decreased degree of hydrogen-bonding in the protein films.

β-Lactoglobulin

Infrared spectra of β-lactoglobulin in different environments are presented in figure 5. The Amide I

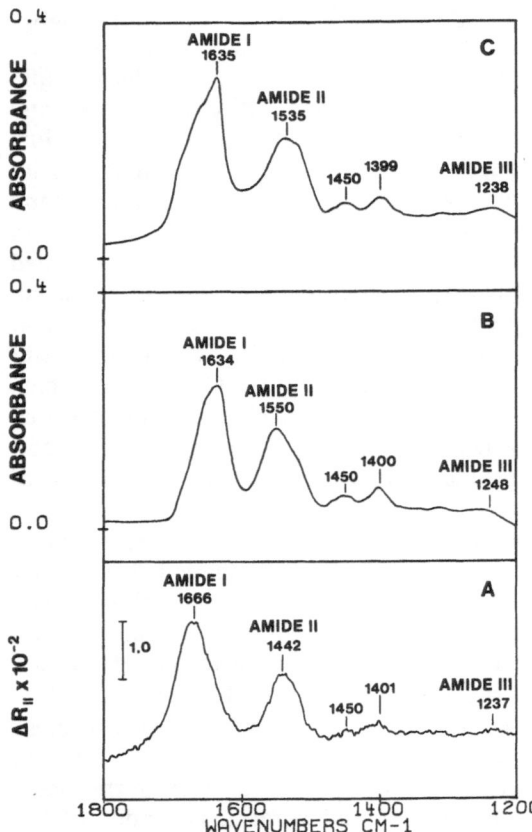

Fig. 5. Infrared spectra of β-lactoglobulin in different environments: (A) IRAS spectrum on gold obtained at 78° and with three reflections; (B) absorbance spectrum in solution; (C) absorbance spectrum of a cast film on a KRS-5 substrate

band of β-lactoglobulin occurs at 1634 cm^{-1} in H$_2$O (pH = 7.0) and at 1635 cm^{-1} in the solid phase. Susi et al. [25] observed this band at 1632 cm^{-1} in D$_2$O, H$_2$O as well as in the solid phase. They assigned this very low Amide I frequency to the antiparallel pleated sheet conformation. Also evident in figures 5B and 5C is the asymmetric shape of the Amide I band. The shoulders on the high frequency side of the Amide I band near 1650 cm^{-1} in both the solution and solid phase spectra are indicative of some disordered content in the protein. The Amide II and Amide III bands which both have a pronounced N−H bending character are as expected very sensitive to changes in the degree of hydrogen-bonding. Blue shifts of 15 and 10 cm^{-1} are observed upon dissolution for the Amide II and Amide III bands respectively.

The Amide I and Amide II bands appear at exactly the same frequencies in the IRAS spectra of β-lactoglobulin on gold and titanium, figure 5A. The bands

occur at 1666 and 1542 cm^{-1} respectively. Accordingly, the blue shifts observed for the Amide I bands of β-lactoglobulin are even larger than those observed for fibrinogen, lysozyme and pepsin on the same metals. It therefore seems reasonable to assume that proteins with dominating β-conformation will undergo a greater change in structure, upon adsorption on metal surfaces, than those with a large amount of α-helix or disordered conformation. It is also observed by comparing figure 5A with figure 5C that the pronounced unsymmetric shape of the Amide I band is no longer present in the IRAS spectrum, indicating that essentially all carbonyl groups in the peptide link experience the same environment on metals. This might suggest that β-lactoglobulin is present as a well orientated film on gold and titanium. It is, however, not possible to assign these relatively high Amide I frequencies on metals to any known chain conformation of β-lactoglobulin in native environments. The interpretation of protein-metal spectra is, as mentioned before, complicated by the possibility that other interactions, such as charge transfer, may contribute to the blue shift as well.

Surface concentration data show that the effective thickness of β-lactoglobulin on gold is approximately 15 Å. A reduced intensity by a factor of two of the Amide II bands in the titanium spectrum indicates that even thinner films are formed on this surface.

Glycine

Glycine in the zwitterionic form exhibits a few characteristic bands in the infrared spectra due to the NH$_3^+$ and CO$_2^-$ groups. The assignments of the vibrational modes corresponding to these groups are made on the basis of infrared spectra of normal and variously deuterated glycine [30, 31] and of ^{15}N- and ^{18}O-labelled glycine [32, 33]. A normal absorption spectrum of glycine in a KBr-matrix is shown in figure 6B. The highest fundamental in this spectrum is the asymmetric stretching vibration of the NH$_3^+$ group at 3167 cm^{-1}. The corresponding symmetric mode is not as easily identified in this spectrum. However, most amino acids exhibit a number of broad bands in the frequency range from 3200 to 2400 cm^{-1}, where at least one of these bands is due to the symmetric NH$_3^+$ stretching mode. The broad absorption between 3200 and 2400 cm^{-1} is also characteristic of the strong hydrogen-bonding existing in the zwitterionic form between neighbouring NH$_3^+$ and CO$_2^-$ groups. The asymmetric and symmetric deformation vibration of the NH$_3^+$ group occurs at 1610 and 1522 cm^{-1} respectively. The ionized CO$_2^-$ group is characterized by two

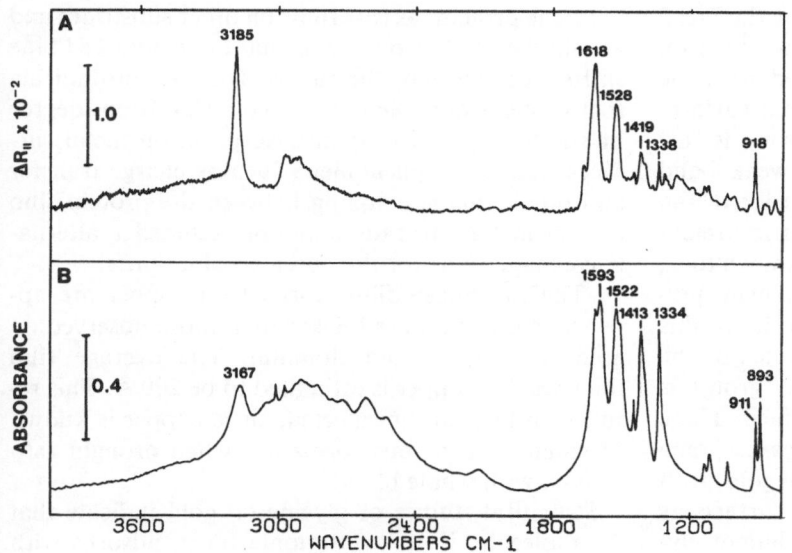

Fig. 6. (A) A single reflection IRAS spectrum of glycine on gold, obtained at 78°. (B) an absorbance spectrum of glycine dispersed in a KBr-matrix and pressed to a pellet

strong bands at 1593 and 1413 cm^{-1} respectively, where the former is the asymmetric stretching vibration which strongly interferes with the asymmetric NH$_3^+$ vibration. The intensity in the 1600 cm^{-1} region is, however, essentially due to the CO$_2^-$ group. This is concluded from infrared data of normal and N-deuterated glycine (ND$_3^+$CH$_2$CO$_2^-$) in KBr. Laulicht et al. [33] showed from infrared spectra of normal and ^{18}O-labelled glycine that the band at 893 cm^{-1} should be assigned to the C^{16}O$_2^-$ scissoring vibration. They also assigned the 1334 cm^{-1} band to the C–C stretching vibration of the glycine backbone.

The reflection spectrum of glycine adsorbed on gold is shown in figure 6A. Some general observations can be made on this figure. The strong and broad absorption in the frequency range from 3200 to 2400 cm^{-1} is completely absent in the reflection spec-

trum, indicating that the strong hydrogen-bonding between neighbouring NH$_3^+$ and CO$_2^-$ groups is drastically reduced upon adsorption on metal surfaces. All bands indicated in figure 6A are shifted towards higher frequencies upon adsorption. This behaviour of these absorption bands is also indicative of a decreased degree of hydrogen-bonding in the adsorbed film. These observations suggest that a preferentially oriented film of glycine is formed on the gold surface. Further evidence for a preferentially oriented film is the large relative change in intensity of some bands in the reflection spectrum compared to the spectrum of a randomly oriented sample in KBr and H$_2$O. The intensities of some characteristic absorption bands of glycine are summarized in table 2. The solution spectrum was obtained at pH = 7.4. The symmetric stretching vibration at 1419 cm^{-1} and the scissoring deformation vibration at 893 cm^{-1} are both associated with a change in dipole moment parallel to the symmetry-axis (the C$_2$-axis of an isolated C$_{2v}$ point group) of the CO$_2^-$ group. According to the "surface selection rule", the reduced intensity of these bands in the IRAS spectrum suggests that the glycine molecule is orientated with the C$_2$-axis mainly parallel to the metal surface. The direction of the dipole moment change of the C–C stretching vibration at 1338 cm^{-1} is essentially parallel to the C–C bond, which in glycine coincides with C$_2$-axis of the CO$_2^-$ group. Accordingly, this band is, as expected, also considerably reduced in intensity upon adsorption on metal surfaces. The symmetric deformation vibration of the NH$_3^+$ group at 1528 cm^{-1} gives rise to a change in dipole moment parallel to the symmetry axis (the

Table 2. Observed intensities of some characteristic bands of glycine in various physical states

Vibrational mode (cm^{-1})[a]	Physical state		
	IRAS	KBr	H$_2$O
ν_{as} (CO$_2^-$) 1593	S	S	S
ν_s (CO$_2^-$) 1413	W	M	M
δ_{sc} (CO$_2^-$) 893	VW	M	–[b]
δ_s (NH$_3^+$) 1522	M	S	M
ν (C–C) 1334	W	M	M

S = strong, M = medium, W = weak, VW = very weak
[a] Refer to KBr frequencies
[b] The cell window has cut-off at 1000 cm^{-1}

C_3-axis of the isolated C_{3v} point group) of the NH_3^+ group. The strong intensity in the 1530 cm^{-1} region indicates that the NH_3^+ group is oriented with the symmetry axis mainly normal to the metal surface. The suggested orientation of the NH_3^+ group is further confirmed by the presence of the weak band around 1035 cm^{-1}, which belongs to the stretching vibration of the $C-N$ bond. The asymmetric stretching vibration of the CO_2^- at 1618 cm^{-1} is known to be associated with a change in dipole moment perpendicular to the C_2-axis. The strong intensity of this band suggests therefore that the plane formed by the carbon and oxygen atoms in the CO_2^- group is oriented mainly normal to the metal surface. These results together indicate that glycine in its zwitterionic form adsorbs in a well defined structure on gold, with the NH_3^+ group pointing towards the surface as shown in figure 7. Angle-dependent XPS data of glycine adsorbed from ethanol solution onto HOPG (highly oriented pyrolytic graphite) substrates [34] confirm the proposed structure in figure 7.

Conclusions

The combination of infrared reflection absorption spectroscopy and the Fourier transform technique has proven to be useful for structural studies of monomolecular protein and amino acid films formed on metal surfaces. All protein films investigated exhibit a distinct blue shift of the Amide I frequency upon adsorption on metal surfaces, compared to the same band in aqueous solution. The magnitude of this blue shift seems to be larger for proteins with dominating β-structure (32 cm^{-1}) than for those with a large amount of α-helix or disordered structures (ca. 20 cm^{-1}). It is concluded from reference spectra of

the same proteins as cast films on other substrates and as multilayer films on metals that these observed blue shifts are induced by the metals. These results indicate that proteins are able to undergo a significant degree of structural change upon adsorption on metal surfaces. However, phenomena such as charge transfer and dipole-dipole coupling between the protein film and the metal substrate cannot be excluded as alternative explanations for the observed blue shifts.

The fibrinogen films formed on copper are approximately ten times thicker than those observed on gold, aluminium and titanium. The average film thickness on copper is estimated to be 200 Å. This result is in fact not unexpected, since copper is known to enhance thrombus formation when brought into contact with whole blood.

Structural studies of glycine on gold indicate that the molecule, in its zwitterionic form, adsorbs with the NH_3^+ group pointing towards the metal surface. This is concluded from the large intensity differences of some CO_2^- and NH_3^+ related bands in the IRAS-spectrum, compared to the same bands in standard transmission spectra of randomly orientated samples in KBr or aqueous solution. The suggested structure on gold is in good agreement with the proposed structure of glycine on HOPG substrates.

Acknowledgements

The studies of proteins and amino acids on solid surfaces are supported by grants from the Swedish Natural Science Research Council and from the National Swedish Board for Technical Development. We wish to thank Mr. Tommy Nylander for stimulating discussions in connection with the β-lactoglobulin experiments.

References

1. Baier RE, Dutton RC (1969) J Biomed Mater Res 3:191
2. Brash JL, Lyman DJ (1969) J Biomed Mater Res 3:175
3. Jacobsen RJ, Brown LL, Winters S, Gendreau RM (1983) J Biomed Mater Res 16:199
4. Gendreau RM, Winters S, Leininger RI, Fink D, Hassler CR, Jacobsen RJ (1981) Appl Spectrosc 35:353
5. Brash JL (1981) In: Salzman EW (ed) Interaction of the Blood with Natural and Artificial Surfaces. Marcel Dekker Inc, New York, p 37
6. Baier RE (1977) Ann N Y Acad Sci 283:17
7. Vroman L, Adams AL, Klings M, Fischer GC, Munoz PC, Solensky RP (1977) Ann N Y Acad Sci 283:65
8. Williams DF (1981) In: Williams DF (ed) Biocompatibility of Clinical Implant Materials, vol 1. CRC Series in Biocompatibility, CRC Press Inc, Boca Raton, Florida, p 9
9. Griss P, Heimke G (1981) In: Williams DF (ed) Biocompatibility of Clinical Implant Materials, vol 1. CRC Series in Biocompatibility, CRC Press Inc, Boca Raton, Florida, p 155

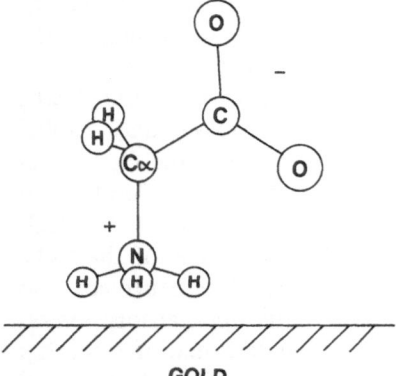

Fig. 7. The suggested structure of glycine adsorbed on gold

10. Francis SA, Ellison AH (1959) J Opt Soc Amer 49:131
11. Greenler RG (1969) J Chem Phys 50:1963
12. Poling GW (1967) J Electrochem Soc 114:1209
13. Boerio FJ, Chen SL (1980) J Colloid Interface Sci 73:176
14. Low MJD, Brown KH, Inoue H (1967) J Colloid Interface Sci 24:252
15. Tompkins HG, Allara DL (1974) J Colloid Interface Sci 49:410
16. Pons SB, Davidson T, Bewick A (1982) J Electroanal Chem 140:211
17. Davidson T, Pons SB, Bewick A, Schmidt PP (1981) J Electroanal Chem 125:237
18. Kottke ML, Greenler RG, Tompkins HG (1972) Surface Sci 32:231
19. Tompkins HG, Greenler RG (1971) Surface Sci 28:194
20. Ortega A, Hoffman FM, Bradshaw AM (1982) Surface Sci 119:79
21. Hollins P, Pritchard J (1980) In: Willis RF (ed) Vibrational Spectroscopy of Adsorbates. Springer-Verlag, Berlin Heidelberg New York, p 125
22. Ito M, Suëtaka W (1978) J Catalysis 54:13
23. Ito M, Mori Y, Kato T, Suëtaka W (1979) Appl Surface Sci 2:543
24. Liedberg B, Ivarsson B, Lundström I (1984) J Biochem Biophys Methods 9:233
25. Susi H (1969) In: Timasheff SN, Fasman GD (eds) Structure and Stability of Biological Macromolecules, Chap. 7. Biological Macromolecules Series 2, Marcel Dekker Inc, New York, p 575
26. Susi H, Timasheff SN, Stevens L (1967) J Biol Chem 242:5460
27. Krimm S (1962) J Mol Biol 4:528
28. Koenig JL, Tabb DL (1979) In: Durig JR (ed) Analytical Applications of FT-IR to Molecular and Biological Systems. Nato Advanced Study Institutes Series C57. D. Riedel Publishing Company, Dordrecht Boston London, p 241
29. Gileadi E, Stanczewski A, Parmeggiani A, Lucas T, Ranganathan M, Srinivasan S, Sawyer PN (1972) J Biomed Mater Res 6:489
30. Gore RC, Barnes RB, Petersen E (1949) Analytical Chem 21:382
31. Pearson JF, Slifkin MA (1972) Spectrochim Acta 28A:2403
32. Tsuboi M, Takenishi T, Nakamura A (1963) Spectrochim Acta 19:271
33. Laulicht I, Pinchas S, Samuel D, Wasserman I (1965) J Phys Chem 70:2719
34. Salaneck WR et al. (1985) Progr Colloid & Polymer Sci 70:83

Received June 27, 1984;
accepted October 15, 1984

Authors' address:

B. Liedberg
Dept. of Physics and Meas. Techn.
Linköping University
S-58183 Linköping (Sweden)

Models of protein adsorption on solid surfaces

I. Lundström

Laboratory of Applied Physics, Linköping Institute of Technology, Linköping (Sweden)

Abstract: A dynamic model of protein adsorption on solid surfaces is presented. It is based on the assumption that a protein molecule may change conformation after adsorption. The computed adsorption isotherms are similar in form to experimentally observed isotherms. A model for the experimentally observed reversible potential changes at a metal electrode upon protein adsorption/desorption is also discussed. It is thus suggested that a second loosely bound protein layer may be present on several metal surfaces. Some notes are also made about the significance of the time scale and detailed method of performing an adsorption experiment.

Key words: Protein adsorption, dynamic model, isotherms, kinetics, potential changes.

Introduction

The adsorption of proteins on solid surfaces is an important phenomenon taking place as soon as a foreign material is brought into contact with a biological system. It is thus involved in situations of bio- and blood compatibility and in fouling in the process industry. Furthermore, there are several surface orientated diagnostic methods based on protein adsorption and interaction on solid surfaces. Proteins are large and complicated molecules and the adsorption process is therefore far from simple to model.

Many adsorption processes are treated thermodynamically in terms of Langmuir type of isotherms where different kinds of interactions between the surface and the molecules and the molecules themselves are incorporated. The adsorption of protein molecules is, however, often a highly dynamic phenomenon. The molecules may change orientation and conformation during or after the adsorption. The properties of the surface play an important role. Protein molecules are normally more influenced by a nonionic or hydrophobic surface than by a polar and hydrophilic surface. We have started to develop simple *dynamic* models for protein adsorption based on the geometrical area covered by a protein molecule in different states [1]. These models can also be extended to incorporate an energetic and geometric interaction between the adsorbed molecules [2].

There are several general experimental observations which should be explained by any theory of protein adsorption on solid surfaces. They are briefly summarized below. Protein adsorption appears to be mainly irreversible. Proteins form generally thin layers on surfaces like gold and platinum and thicker more extended layers on (oxidized) metals like titanium and zirconium. Thus the substrate surface free energy determines to some extent the structure of the adsorbed protein layer. The adsorption isotherms often show a step or plateau at low protein concentrations, followed by a slow increase in the number of adsorbed molecules over a large range in protein concentration. Recent studies are found in references [3 – 8]. Some observations of kinetic nature are briefly described below [3, 8]. If, e.g., a protein is adsorbed on a hydrophobic silicon surface, the amount of adsorbed protein depends on how the protein is added to the solution. If the protein is added in steps at certain time intervals, the adsorbed amount becomes smaller than if the final concentration is added at once [8]. This effect depends both on the nature of the protein and the surface. It is generally smaller on a hydrophilic (silicon) surface [8]. When protein molecules adsorb on a metal surface, there is a change in the potential, which occurs mainly after the protein molecules have adsorbed on the surface. The potential change often contains a large reversible part, which is observed upon rinsing of the surface, although most of the adsorbed protein molecules stay on the surface [9, 10]. A further observation is that "denaturated" protein molecules can be found in the solution after certain adsorption experiments [3]. This indicates that protein molecules do come off the surface with a changed conformation. There appears to be a contradiction between this observation and the fact that protein molecules by and large are irreversibly adsorbed. It is, however, easy to show by thermodynamic arguments that exchange of molecules on the surface may be much more likely than a spontaneous desorption in the absence of other molecules. Exchange reactions are therefore possible and do occur. They will be treated of in a separate communication.

If we return to the adsorption isotherms and the adsorption kinetics, it is obvious that they cannot be explained by a simple Langmuir type of adsorption. One question is, of course, if adsorption from a heterogeneous protein solution together with different forms of interaction on the surface can give a modified Langmuir adsorption isotherm and kinetics, which may account for the experimental results. We have tried several models based on Langmuir-like behaviour, and found that none of them can explain all the facts of the behaviour of protein adsorption on solid surfaces. We limit ourselves to discuss some general features of Langmuir adsorption below.

We believe that a realistic model for protein adsorption should contain a description of the dynamics of protein molecules adsorbed on surfaces, as suggested earlier [1]. We therefore develop this dynamic model further in the following.

Finally, we use the dynamic model to explain the (reversible) potential changes observed experimentally [9, 10]. We suggest that in several cases, a loosely bound protein layer develops on the first (irreversibly) adsorbed layer. Further experimental results and a discussion of the possible nature of the second protein layer are found in reference [10].

Simple dynamic model for protein adsorption

We will examine the case where a protein molecule is adsorbed in one conformation and then undergoes an irreversible conformational change on the surface (fig. 1a). There are, of course, several other possibilities as indicated by the situation in figure 1b. We start with a certain *desorption* from both of the adsorbed states. We make the assumption that a protein molecule covers an area on the surface of A_1 and A_2 in the two states, respectively. We define the maximum possible number of adsorbed protein molecules per unit area, N_0, when the surface is covered only by molecules in state 1, by letting $N_0 A_1 = 1$.

We introduce the coverages of molecules adsorbed in state 1 and 2 respectively

$$\theta_1 \equiv n_1/N_0 \quad \text{and} \quad \theta_2 \equiv n_2/N_0$$

where n_i is the number of adsorbed molecules (per unit area). Note that the geometrical coverages are $N_0 A_1 \theta_1$ and $N_0 A_2 \theta_2$ respectively. If the protein concentration (of type 1) in the solution is C_1 and with the rate constants defined by the solid arrows in figure 1a, we obtain the kinetic equations

$$\frac{d\theta_1}{dt} = (k_1 C_1 - s_1 \theta_1)(1 - \theta_1 - a\theta_2) - r_1 \theta_1 \tag{1}$$

$$\frac{d\theta_2}{dt} = s_1 \theta_1 (1 - \theta_1 - a\theta_2) - r_2 \theta_2 \tag{2}$$

where $a \equiv A_2/A_1$. Equations (1) and (2) yield in steady state

$$\theta_1 = \left[\left(\frac{k_1 C_1 + s_1 + r_1}{2(ar_1/r_2 - 1)s_1} \right)^2 + \frac{k_1 C_1}{(ar_1/r_2 - 1)s_1} \right]^{1/2}$$
$$- \frac{k_1 C_1 + s_1 + r_1}{2(ar_1/r_2 - 1)s_1} \tag{3}$$

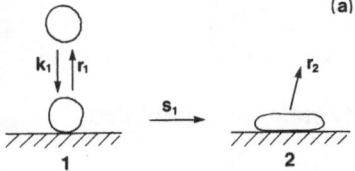

Fig. 1(a). Schematic illustration of the adsorption of a protein molecule undergoing conformational changes on the surface. The rate constants for the different processes are defined in the drawing

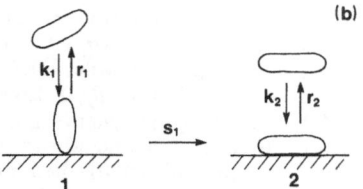

Fig. 1(b). Schematic illustration of the adsorption of a protein molecule, which can be adsorbed with two different orientations

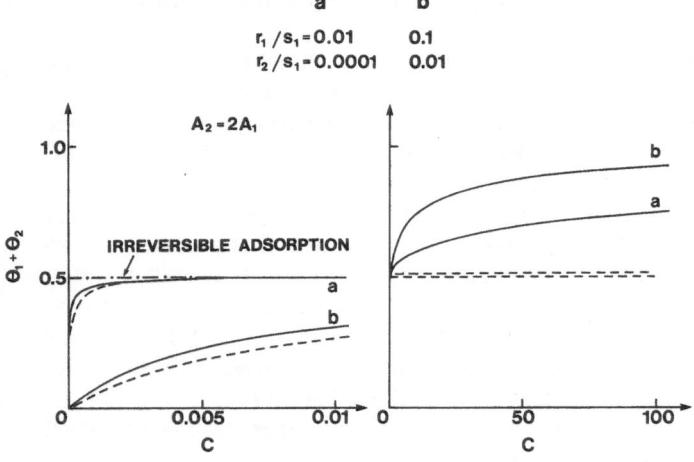

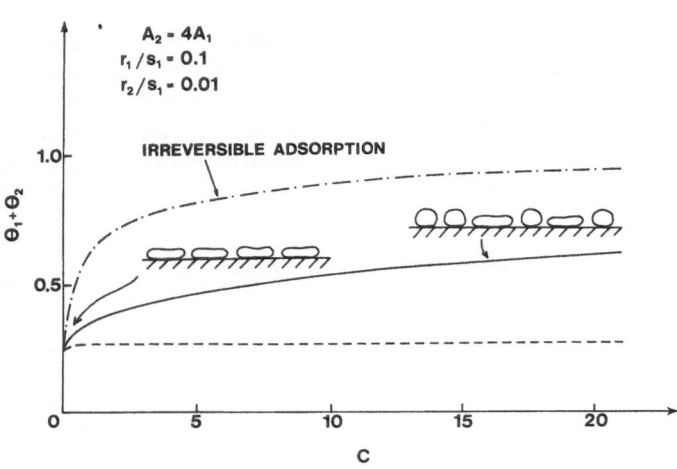

Fig. 2. Upper diagram: Illustration of the basic features of the dynamic model at low and high protein concentrations, respectively. Some different desorption time constants were used. C is the normalized concentration $C \equiv k_1 C_1/s_1$. Langmuir-like adsorption of a molecule in two different orientations is denoted by the dashed lines. The parameters were chosen to obtain a good fit between the Langmuir case and the dynamic model at low concentrations. It is also indicated that an irreversible adsorption (dashed-dotted line) would lead to a constant adsorption at low concentrations equal to $\theta_1 + \theta_2 = 1/a$ for the dynamic model. Note that $(\theta_1 + \theta_2)$ corresponds to the number of adsorbed protein molecules/per unit area. (The number is actually $N_0(\theta_1 + \theta_2)$.) Lower diagram: $\theta_1 + \theta_2$ for $A_2 = 4A_1$ comparing the dynamic model with reversibility (solid line) with an irreversible adsorption (dashed-dotted line) and a Langmuir isotherm (dashed line) at intermediate concentrations. The irreversible adsorption is only approximate, given by equations (9) and (10). We have also shown schematically the packing of the protein molecules in different concentration regions

and

$$\theta_2 = \frac{s_1 r_1 \theta_1^2}{r_2(k_1 C_1 - s_1 \theta_1)} \tag{4}$$

which give adsorption isotherms for $\theta_1 + \theta_2$ different in form from a normal Langmuir isotherm as indicated in figure 2. They are actually very similar to the experimental isotherms observed. Note that $\theta_1 + \theta_2$ corresponds to the number of adsorbed protein molecules and hence to the adsorbed amount of organic material on the surface. The Langmuir isotherm was obtained by assuming that we have two kinds of protein molecules in the solution, one adsorbing covering an area A_1 and one covering an area A_2 but with no conformational change on the surface. The resulting Langmuir-like isotherm for $\theta_1 + \theta_2$ is then

$$\theta_1 + \theta_2 = \frac{1 + \dfrac{k_2 C_2 r_1}{k_1 C_1 r_2}}{1 + a \dfrac{k_2 C_2 r_1}{k_1 C_1 r_2}} \quad \frac{C_1}{C_1 + \dfrac{r_1/k_1}{1 + \dfrac{ak_2 C_2 r_1}{k_1 C_1 r_2}}} . \tag{5}$$

The same expression with $C_1 = C_2$ is obtained if the protein molecule can be adsorbed directly from solution in two different conformations and with no change of conformation on the surface. This is the situation which is most comparable with the simple dynamic model.

We have treated a situation with reversible adsorption. By making r_1 and r_2 very small, we can in practice mimic irreversible adsorption.

For true irreversible adsorption $r_1 = r_2 = 0$, we have to use a kinetic argument to obtain θ_1 and θ_2 [1]. From equations (1) and (2) we get

$$\frac{d\theta_2}{d\theta_1} = \frac{s_1 \theta_1}{k_1 C_1 - s_1 \theta_1} \tag{6}$$

and with θ_1 and $\theta_2 = 0$ at $t = 0$,

$$\theta_2 = -\theta_1 - \frac{k_1 C_1}{s_1} \ln\left(1 - \frac{\theta_1 s_1}{k_1 C_1}\right). \tag{7}$$

Furthermore, in steady state we require that

$$1 - \theta_1 - a\theta_2 = 0 . \tag{8}$$

These two equations define θ_1 and θ_2 which have to be determined by computer, iteration or through the use of suitable diagrams [1]. We discuss here an approximate solution for $\theta_1 s_1/k_1 C_1 \ll 1$. In this case

$$\theta_2 \approx \frac{s_1 \theta_1^2}{2 k_1 C_1} \tag{9}$$

and

$$\theta_1 \approx \left[\left(\frac{k_1 C_1}{s_1 a}\right)^2 + \frac{2 k_1 C_1}{s_1 a}\right]^{1/2} - \frac{k_1 C_1}{s_1 a}. \tag{10}$$

$\theta_1 + \theta_2$ according to this approximation is also plotted in figure 2. Even if the approximation is not good for small Cs it indicates that the adsorption isotherms have the same form for irreversible and reversible adsorption in the presented model. The adsorbed amount will naturally be smaller in the reversible case. An important parameter is then the relative magnitude of the desorption rate constant for the two types of protein molecules adsorbed on the surface. A small r_2/r_1 means that relatively more molecules of type 2 will be found on the surface at large concentrations.

It is in this context of interest to discuss the irreversible Langmuir case

$$\frac{d\theta_1}{dt} = k_1 C_1 e^{\alpha_1 (\theta_1 + a\theta_2)} (1 - \theta_1 - a\theta_2) \tag{11}$$

$$\frac{d\theta_2}{dt} = k_2 C_2 e^{\alpha_2 (\theta_1 + a\theta_2)} (1 - \theta_1 - a\theta_2) \tag{12}$$

where we have also introduced coverage dependent rate constants. In this case, we find that (if $\theta_1 = \theta_2 = 0$ at $t = 0$)

$$e^{-\Delta \alpha a \theta_2} = 1 + \frac{k_2 C_2}{k_1 C_1} a - \frac{k_2 C_2}{k_1 C_1} a e^{\Delta \alpha \theta_1} \tag{13}$$

and equation (8) determine the coverages θ_1 and θ_2. ($\Delta \alpha = \alpha_2 - \alpha_1$). It is particularly interesting to note that equation (13) depends only on C_2/C_1. To obtain an adsorption isotherm, experiments with different concentrations of a protein are carried out. The ratio C_2/C_1 ought, however, to be a constant for a given protein stock solution. This means that the adsorption isotherm is a straight line. In the special case that $\alpha_1 - \alpha_2 = 0$, we obtain

$$\theta_1 = \frac{1}{1 + a\dfrac{k_2 C_2}{k_1 C_1}} \tag{14}$$

$$\theta_2 = \frac{k_2 C_2/k_1 C_1}{1 + a k_2 C_2/k_1 C_1} \tag{15}$$

and

$$\theta_1 + \theta_2 = \frac{1 + k_2 C_2/k_1 C_1}{1 + a k_2 C_2/k_1 C_1}, \tag{16}$$

i.e. independent of the total concentration $C_1 + C_2$ as long as C_2/C_1 stays constant. Hence, irreversible Langmuir-like adsorption into two different conformations cannot explain the form of the observed adsorption isotherms.

The basic features of the present model are illustrated in figure 2. At low concentrations, a thin layer is formed mainly by molecules in state 2. At very high concentrations, the surface is covered mainly by molecules in state 1 since there is very little time for the molecules to change conformation before the surface is covered. We can say that a close packed layer of native molecules has formed. In the irreversible case $\theta_1 + a\theta_2 = 1$ in steady state. This means that in an experiment with subsequent addition, the first addition will determine the amount of adsorbed protein molecules since when $\theta_1 + a\theta_2 = 0$ no area exists for adsorption. In experiments, an increase of adsorption can also be seen at subsequent additions, although in many cases the final value becomes smaller than on direct addition of the final concentration [3, 8] (as illustrated in fig. 3). In a full account of this type of experiment, the time scale of the experiment, exchange reactions, multiconformational states, and a possible small reversibility become important. These factors will be discussed in a separate paper.

The model above was developed for a conformational change of the protein molecule after adsorption. It also works for a nonglobular protein molecule which changes its orientation on the surface after adsorption. If a protein molecule is also allowed to adsorb directly with the "second" orientation (corresponding to fig. 1b) we obtain a slight modification of the adsorption isotherms. With reversibility we obtain

$$\theta_1 = \left[\left(\frac{(k_1 + k_2 a r_1/r_2) C_0 + s_1 + r_1}{2(a r_1/r_2 - 1) s_1}\right)^2 \right.$$
$$\left. + \frac{k_1 C_0}{(a r_1/r_2 - 1) s_1}\right]^{1/2}$$
$$- \frac{(k_1 + k_2 a r_1/r_2) C_0 + s_1 + r_1}{2(a r_1/r_2 - 1) s_1} \tag{17}$$

and

$$\theta_2 = \frac{r_1}{r_2} \theta_1 (k_2 C_0 + s_1 \theta_1)/(k_1 C_0 - s_1 \theta_1). \tag{18}$$

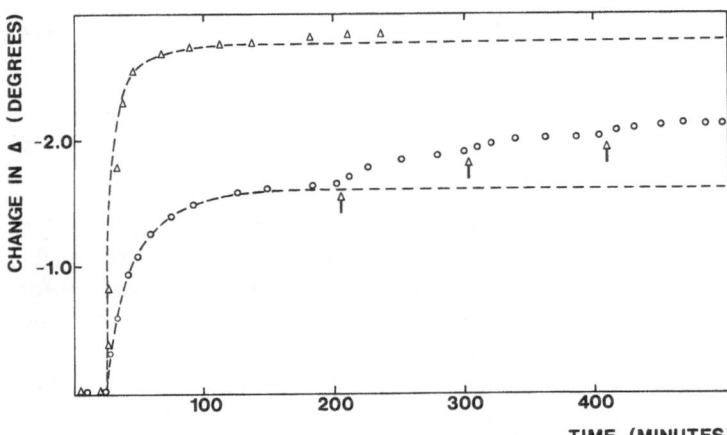

Fig. 3. Adsorption kinetics for human fibronectin on a hydrophobic silicon surface. The parameter "Delta" is obtained from ellipsometric studies and is a measure of the adsorbed amount of organic material on the surface. The triangles represent the result when the final protein concentration (30 µg/ml) was added at once. The circles represent subsequent additions to a final concentration of 30 µg/ml. The first addition was 2.5 µg/ml. (The dashed lines show the type of kinetics expected from the simple dynamic model presented here if steady state is achieved during the first addition.) The experimental points are taken from Jönsson et al. [8]

For the irreversible case we obtain

$$\theta_2 = -\theta_1 - \frac{k_2 + k_1}{s_1} C_0 \ln \left(1 - \frac{s_1 \theta_1}{k_1 C_0} \right) \qquad (19)$$

together with equation (8) as before.

C_0 is the protein concentration in solution.

Reversible potential changes

The adsorption of protein molecules on metal surfaces can be followed *in situ* by, e.g., ellipsometry [9, 11, 12], capacitance [9, 13] and surface potential measurements [9]. From ellipsometric measurements, the (optical) mass of the organic layer can be deduced, capacitance measurements probably yield information about the number of contact points or the contact area between the protein molecule and the surface. It is natural to assume that the potential changes would follow the capacitance changes, if their origin is due to the displacement of water molecules and charges from the surface. In recent simultaneous ellipsometric, capacitance and potential measurements, we have observed that, whereas the ellipsometric and capacitance measurements indicate an irreversible adsorption, the potential change often contains a large reversible part which disappears upon rinsing [9, 10]. Most of the potential change also occurs at the end of the adsorption kinetics, as demonstrated in figure 4a.

The size and reversibility of the potential change is both protein and substrate dependent. We have suggested that the reversible part of the potential change is due to a second protein layer loosely associated with the first (irreversibly) adsorbed layer [10]. This loosely associated layer does not in general yield any large ellipsometric effects. It is assumed to be hydrated with a refractive index close to that of water. If we assume that the second layer is adsorbed in a (reversible) Langmuir fashion mainly on protein molecules of type 2, we obtain the following kinetic equation for the coverage of the second layer, θ_s,

$$\frac{d\theta_s}{dt} = k_s C_1 (\theta_2 - \theta_s) - r_s \theta_s . \qquad (20)$$

We have used a computer program for coupled first order differential equations to calculate the time dependencies of θ_1, θ_2 and θ_s during an adsorption experiment. This program, which is developed for prob-

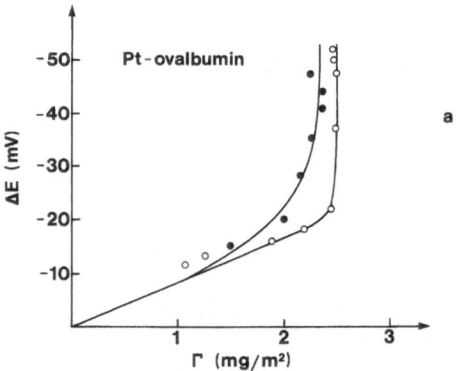

Fig. 4(a). Experimental results from simultaneous ellipsometric and potential measurements during protein adsorption (for ovalbumin on platinum). The surface potential change at a given time is plotted against the mass of the organic layer on the surface at the same time. The drawing indicates that most of the potential change occurs at the end of the adsorption process. The two curves represent two different experiments under identical conditions.

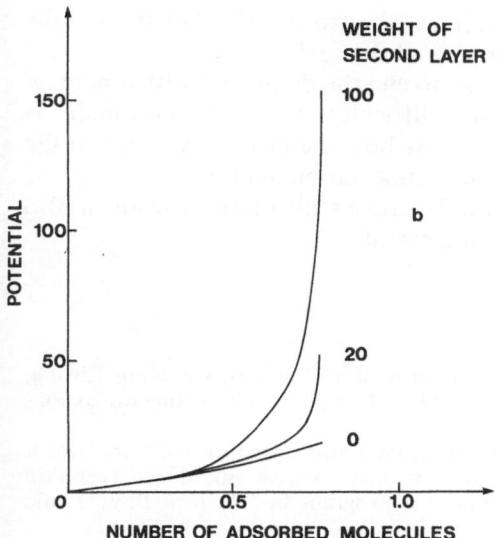

Fig. 4(b). Theoretically calculated potential changes (eq. (21)) with $\Delta E_0 = 15$, $p_2 = 4$ and p_s given by the indicated numbers. The x-axis is the normalized number of adsorbed molecules, $\theta_1 + \theta_2$. The time dependencies of θ_1 and θ_2 were calculated (from eqs. (1) and (2)) with the following parameters: $r_1 = r_2 = 0$ (irreversible adsorption); $A_2 = 2A_1$ and $C \equiv k_1 C_1/s_1 = 1$ with $\theta_1 = \theta_2 = 0$ at $t = 0$. θ_s was calculated from equation (20) with $k_s C_1/s_1 = 0.2$ and $r_s/s_1 = 0.1$ (in the calculations, the time, τ, was normalized with respect to s_1, i.e., $\tau = s_1 t$)

lems in control engineering, generates the coverages as a function of time. To make a comparison with the experimental results in figure 4a, we have plotted calculated $\theta_1 + \theta_2$ versus an assumed potential change, ΔE of the form

$$\Delta E = \Delta E_0(\theta_1 + p_2\theta_2 + p_s\theta_s) \qquad (21)$$

at given times. $\theta_1 + \theta_2$ corresponds to the number of organic molecules adsorbed in the first layer on the surface (or the ellipsometrically determined mass per unit area, Γ). Figure 4b indicates that if a large potential change is caused by the second layer then the theoretically calculated curve looks very similar to the experimental one.

We have thus at least a mathematical argument for the hypothesis that a potential change can be caused by a second protein layer adsorbed on the first layer. If one agrees with the idea above, there are several questions which have to be answered. Are the potential changes due to polarisation changes at the metal surface, changes in corrosion potential, redox levels etc. induced by the protein? Are the protein molecules in the second protein layer physically different from the molecules in the solution?

A note on the time scale of an adsorption experiment

The computer program mentioned above was used to calculate the time dependencies of θ_1 and θ_2 (from eqs. (1) and (2)) assuming small desorption rate constants r_1 and r_2. The results are shown in figure 5, where they are compared with a true irreversible adsorption. At short times, we get an adsorption close to the irreversible case, which at longer times is followed by a desorption of molecules in state 1 (in the example chosen). It is obvious that the time scale of an experiment is extremely important if there is a slow desorption (or exchange) of molecules on the surface. Furthermore, on the time scale in figure 5(a), the spontaneous desorption upon rinsing will be small, whereas on the time scale in figure 5(b) a notable desorption will occur.

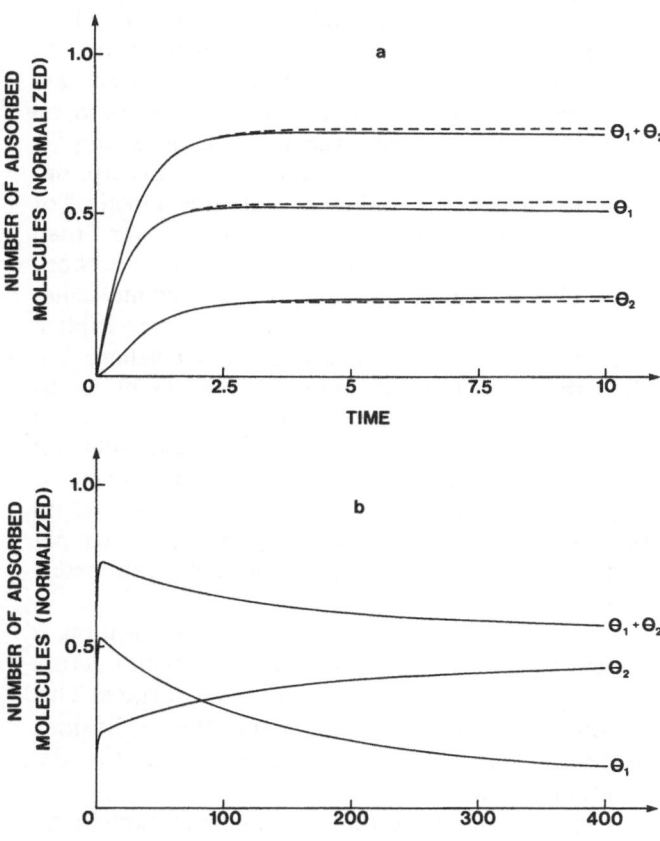

Fig. 5. Calculated transients from equations (1) and (2) of θ_1, θ_2 and $\theta_1 + \theta_2$ during an adsorption experiment for the case when $C = k_1 C_1/s_1 = 1$; $r_1/s_1 = 0.01$; $r_2/s_1 = 0.0001$; $A_2 = 2A_1$. The time scale was normalized with respect to s_1, i.e. Time $= s_1 t$.
(a) Adsorption kinetics at short times. The dashed lines were obtained for true irreversible adsorption ($r_1 = r_2 = 0$).
(b) Adsorption kinetics during a longer time

Discussion

A simple dynamic model for protein adsorption was proposed to account for the possibility of protein molecules changing conformation or orientation on solid surfaces. The relative number of molecules which have undergone a conformational change depends on the concentration in solution and the time constant for the surface induced conformational changes. The model should be extended to incorporate not only two but several conformational states on the surface. Coverage dependent conformations, i.e., interaction between the adsorbed protein molecules, should be included. The possibility of a lateral mobility of the protein molecules on the surface and its influence on the adsorption isotherms ought to be investigated. We have on several occasions stated that the adsorption of proteins is irreversible in many cases. Some experiments indicate that this may be a matter of time scale and not a fundamental fact. In this context exchange reactions on the surface are particularly interesting. It is easy to understand on energetic grounds that exchange reactions can be much more probable than spontaneous desorption. Model calculations indicate that the kinetics of exchange reactions can be much faster than spontaneous desorption. Exchange reactions between molecules in solution and already adsorbed molecules need therefore to be considered before we can explain the fine details of protein adsorption. We believe that the extension of the theoretical model as discussed above can be used to explain, e.g., experimental results like those in figure 3. It is obvious that not only the nature of the protein but also of the surface is important for the behaviour of protein molecules on the surface. Surface energy and polarity, surface pH and redox potential are some of these parameters [14].

In some cases, the rate of protein adsorption is limited by the diffusion of protein molecules across the "unstirred" layer close to the solid surface. This possibility will, however, not change the conclusions made about the adsorption isotherms and slow desorption kinetics.

It is also interesting that our dynamic model may account for the kinetics of the (reversible) potential changes with the assumption of a second loosely bound protein layer. It must be very extended and

hydrated since it gives rise to very small changes in the optical parameters of the surface.

We would like to end the discussion with a note on an experimental difficulty. If a dynamic model is valid then it matters how the protein is added to the solution. The experimental method may therefore be very important. The time scale of an experiment also appears to be important.

Acknowledgements

I am grateful to Prof. Kåre Larsson, Dr. Hans Elwing, Bengt Ivarsson, and Ulf Jönsson for stimulating discussions on the present topic.

Our research on protein adsorption on solid surfaces is supported by the National Swedish Board for Technical Development under a program in "Surface Physics and Chemistry".

References

1. Lundström I (1983) Physica Scripta T4:5
2. Lundström I, Elwing H (1984) J Theor Biol 110:195
3. Walton AG, Soderquist ME (1980) Croatica Chemica Acta 53:363
4. Lyklema J (1980) Croatica Chemica Acta 53:353
5. Fair BD, Jamieson AM (1980) J Colloid Interf Sci 77:524
6. Kim SW, Lee RG (1975) In: Baier RE (ed) Applied Chemistry at Protein Interfaces, Adv Chem Ser 145:218
7. Norde W, Lyklema J (1979) J Colloid Interf Sci 71:350
8. Jönsson U, Ivarsson B, Lundström I, Berghem L (1982) J Colloid Interf Sci 90:148
9. Ivarsson B, Hegg P-O, Lundström I, Jönsson U (1985) Colloids and Surfaces (in press)
10. Arnebrant T, Ivarsson B, Larsson K, Lundström I, Nylander T (1985) Progr Colloid and Polymer Sci 70:62
11. Vroman L, Adams AL, Klings M, Fischer GC, Munoz PC, Solensky RP (1977) Ann N Y Acad Sci 283:65
12. Azzam RMA, Bashara NM (1977) Ellipsometry and Polarized Light. North Holland, Amsterdam
13. Stoner G, Srinivasan S (1970) J Phys Chem 74:1088
14. Hoffman AS (1982) In: Cooper SL, Peppas NA (eds) Biomaterials: Interfacial Phenomena and Applications. Adv Chem Ser 199:3

Received June 30, 1984;
accepted October 30, 1984

Author's address:

Ingemar Lundström
Laboratory of Applied Physics
Linköping Institute of Technology
S-58183 Linköping (Sweden)

Progress in Colloid & Polymer Science   Progr Colloid & Polymer Sci 70:83 – 88 (1985)

Photoelectron spectroscopy of amino acids adsorbed upon surfaces: glycine on graphite

W. R. Salaneck*, I. Lundström**, and B. Liedberg**

*Laboratory of Surface Physics and Chemistry and **Laboratory of Applied Physics, IFM University of Linköping, Linköping (Sweden)

Abstract: The physical adsorption of glycine molecules on highly oriented pyrolytic graphite substrates is studied using angle-dependent X-ray photoelectron spectroscopy, XPS(θ). A partially oriented double-layer is found to be stable on the surface. The structure of the double-layer is consistent with (but not identical to) the crystal structure of β-glycine, which is a layered crystal.

Key words: Amino acids, glycine, graphite, photoelectron spectroscopy

Introduction

The nature of the interactions of proteins with non-biological surfaces is the basis for the biological acceptance or rejection of artificial implants in living systems. A wide variety of *in vivo* as well as model *in vitro* studies are carried out to observe conditions under which proteins adsorb from solution onto non-biological surfaces. Previously, we have carried out a preliminary study of the behavior of one of the smallest amino acids, the building blocks of polypeptides and proteins, on non-biological surfaces, studied under ultra high vacuum conditions using surface-sensitive photoelectron spectroscopy [1].

The focus of the preliminary survey study was to determine whether or not there is any relationship between the behavior of amino acids on the surfaces of non-biological materials, as studied in an ultra high vacuum environment using modern surface science techniques, and the known biological behavior of these same materials *in vivo*. The utilization of small model molecules in conjunction with surface sensitive photoelectron spectroscopies should then allow the elucidation of microscopic details of the biomolecule-surface interaction *albeit* in an artificial environment. We started with glycine, the smallest and most simple amino acid, shown in figure 1. We have found that the general behavior of glycine molecules on diverse materials surfaces can be divided broadly into three categories, each corresponding to a known degree of biocompatibility of these same materials [1].

In this report, we discuss the details of the results of an investigation of the adsorption of glycine mole-cules on the surface of graphite (highly oriented pyrolytic graphite, or HOPG) using a form of angular dependent X-ray photoelectron spectroscopy, XPS(θ), figure 2. XPS (to the physicist) is also known as ESCA, for *e*lectron *s*pectroscopy for *c*hemical *a*pplications, especially to chemists. The biological background for this type of study can be found discussed in reference [1]. It was possible to carry out this study of glycine on HOPG with modest sample

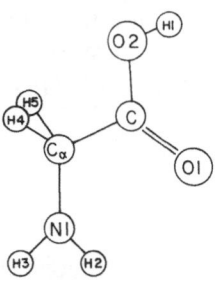

NEUTRAL

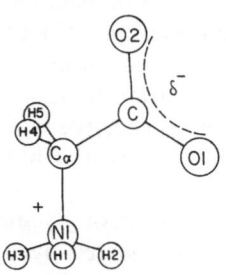

ZWITTERION

Fig. 1. The structure of neutral and zwitterionic glycine [15]

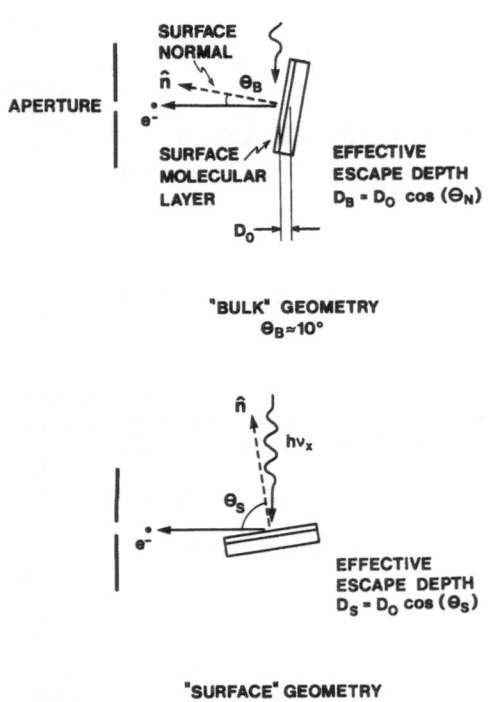

Fig. 2. The geometry used in the XPS(θ) data acquisition. Do is typically taken to be twice λ_e, the elastic-mean-free-path. See text

handling facilities on a rather simple XPS instrument (see next section), because (A) spectroscopically clean and simple HOPG surfaces are easy to prepare reproducibly; and (B) because the sample with a thin glycine over-layer on HOPG are easy to prepare, stable in UHV for long periods of time, and very reproducible.

Experimental results

It seemed appropriate, given the ultimate biological focus of our previous survey study [1], to employ "air-exposed" surfaces, at least initially. For example, typical model physical-chemical studies of protein adsorption on metal surfaces are carried out on air exposed surfaces [2]. Our array of substrate materials were chosen both for their desirable spectroscopic characteristics as well as their real or potential biophysical applications [1].

In this report, we focus on the adsorption of glycine on freshly exposed surfaces of highly oriented pyrolytic graphite, or HOPG. Samples were "tape-cleaved" in air to expose a new surface, and then inserted directly into a Vacuum Generators XPS instru-

ment through a Kratos sample insertion lock mechanism. Spectra were recorded using unfiltered Al k_α X-radiation, under conditions such that the full-width-at-half-maximum of the Au4$f_{7/2}$ peak was 1.3 eV. XPS(θ) core-level and valence band spectra of such surfaces agree well with published XPS spectra of vacuum cleaved HOPG [3]. In particular, no significant oxygen signal was observed, and the $C(1s)$ main line, well known plasmon satellite, and the valence band features were equivalent to the previously published spectra of Wertheim and coworkers [3]. The spectra of clean HOPG are included in figures 3 and 4.

The substrates were then removed from the spectrometer, immediately immersed into a saturated solution of glycine in pure ethanol, and reinserted into the spectrometer to dry. This entire removal and reinsertion operation took typically 2 minutes. The XPS spectra of HOPG, removed for 2 minutes without the application of the glycine solution, remained essen-

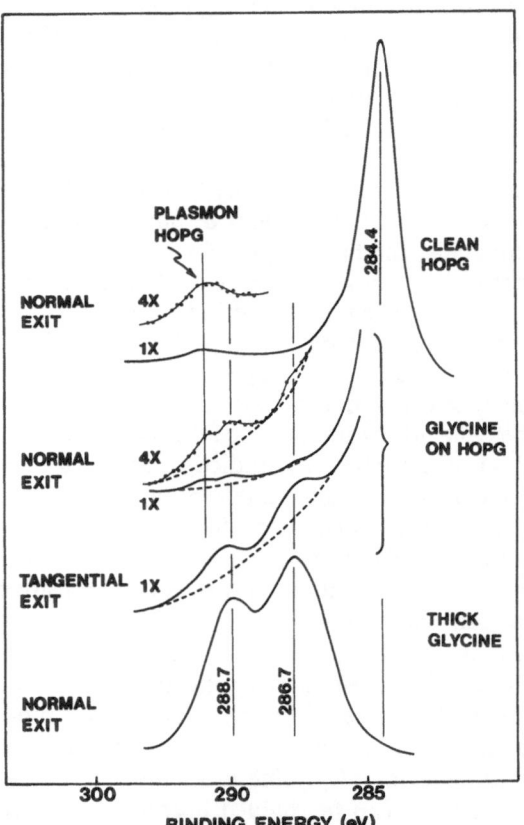

Fig. 3. The $C(1s)$ spectra of clean HOPG compared with that of glycine on HOPG and bulk glycine. The HOPG and bulk glycine spectra are in good agreement with those in references 3 and 4, respectively

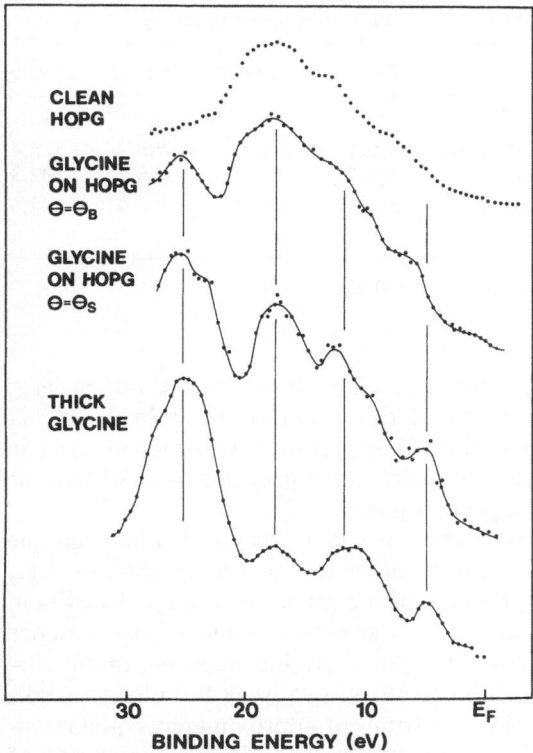

Fig. 4. The XPS(θ) valence band spectra of clean HOPG compared with those of glycine on HOPG and bulk glycine. Note some subtle electronic structural differences between the data for $\theta = \theta_s$ and $\theta = \theta_B$. The spectra for clean HOPG and for bulk glycine compare favorably with those in references 3 and 4, respectively

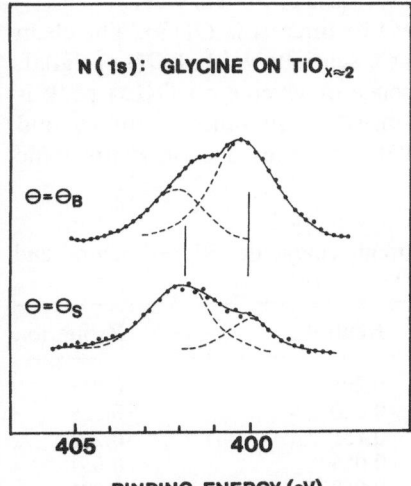

Fig. 5. The $N(1s)$ core-level spectra for glycine on a $TiO_{x \approx 2}$ surface with a weak nitrogen impurity. The spectra show that with XPS(θ) one can distinguish between the two species and observe that the glycine $N(1s)$ at 402 eV lies on top of the impurity

tially unchanged. If the HOPG was immersed into pure ethanol only, the XPS spectra were essentially unchanged upon reinsertion into the spectrometer.

Glycine is very weakly soluble in ethanol, and thus only a thin layer of glycine molecules was left upon the HOPG surface after evaporation of the ethanol. These layers were still very thick compared with a monolayer, however. During about the first two hours in the spectrometer, the overlayers of glycine evaporated, until a stable signal was reached, corresponding to approximately a double molecular layer (see below). Thus the use of an ethanol solution is merely one method for applying a very thin film of glycine onto the HOPG surface. Subsequent thinning of the film occurs by natural evaporation of the glycine overlayers. The remaining (multi)-layer was stable on HOPG at room temperature for up to 2 days (the maximum time observed) in the vacuum chamber ($p \sim 1 \times 10^{-8}$ Torr) and even under continuous Al k_α X-radiation.

The evaporation rate of glycine was estimated by starting with macroscopically thick glycine films formed on various substrates by solution casting methods, then measuring the time in the spectrometer (at $\sim 10^{-8}$ Torr) until a substrate XPS signal could be detected. Such a measurement is estimated to be correct within an order of magnitude. We obtain an evaporation rate of approximately one mono-layer per 10 seconds. The initial film thickness (upon evaporation of the ethanol) must be no more than about 1000 to 2000 Å in order to evaporate in less than 2 hours. Thus, the stability of the final layer(s) of glycine on HOPG for over 2 days is a measure of the significant stability of the adsorbate system. The XPS spectra of thick (>200 Å) layers of glycine are also contained in figures 3 and 4. The core-level binding energies and valence band features are in very good agreement with ESCA spectra previously published by Clark and co-workers [4].

The angle of collection of the photoelectrons relative to the plane of the sample surface was varied by rotating the sample, as shown in figure 2. Observing electrons at near normal exit allows the collection of photoelectrons generated deeper within the sample (with an escape depth, λ, of 50 to 100 Å, depending upon which part of the entire XPS spectrum is involved). Observing electrons near grazing exit emphasizes the very outermost layers of the sample, and suppresses signal from electrons generated deeper within the sample. These θ-dependent effects have been discussed previously in general [5], and for organic and polymeric systems in particular [6, 7]. The effective escape depth varies with angle (fig. 2) approximately as $\lambda_{\mathrm{eff}} = \lambda_0 \cos\theta$. For $\theta = \theta_B = 10°$, $\lambda_{\mathrm{eff}} \approx \lambda_0$ (Bulk sensitive), whereas for $\theta = \theta_s = 75°$, $\lambda_{\mathrm{eff}} \approx 0.25\,\lambda_0$ (surface sensitive). The simple θ-dependent effect can be seen in the $N(1s)$ spectra of the overlayer of glycine on the natural oxide of titanium, $\mathrm{TiO}_{x \approx 2}$, in figure 5. Two things can be learned from the spectra: (1) that the glycine nitrogen at $E_B \approx 402$ eV can be distinguished from a weak nitrogen impurity at $E_B \approx 400$ eV that is sometimes found on $\mathrm{TiO}_{x \approx 2}$; and (2) that the glycine nitrogen lies on top of the impurity nitrogen (as seen for $\theta = \theta_s$). No such N-impurities are detected on HOPG, however. For clean HOPG, the valence band spectrum is rather featureless, as seen in figure 4. Using $\theta = \theta_s$, the valence band features of the thin over-layer of glycine are emphasized over the more or less smooth background. The same is true for the $C(1s)$ core level spectra in figure 3, where the signal from the two glycine carbon atoms, having higher binding energies than the HOPG carbon atoms, can be enhanced for $\theta = \theta_s$. The various atomic binding energies are tabulated in table 1.

Discussion

The electronic structure of glycine, as measured by ultraviolet photoelectron spectroscopy, or UPS, has been studied previously in both the gas phase [8] and the solid phase [9]. As is well known, glycine exists as a neutral molecule in isolation, i.e., in the gas phase [10]. In solution or in the (molecular) solid phase, however, the molecules exist as zwitterions [11], as illustrated in figure 1. In the zwitterionic form, the carboxyl-group is proton deficient and the amino-group has an extra proton, which corresponds to a large measured dipole moment of about 15.7 D [12]. The total energy of the isolated zwitterionic molecule is greater than that of the isolated neutral molecule. In the molecular solid, however, a net energy saving is experienced due to the strong dipole-interactions, which stabilize the zwitterionic form [9]. In other words, the total energy per unit volume of the solid is minimized when glycine is in the zwitterionic form. One manifestation of the existence of the zwitterionic form in the solid phase is the significant change in the UPS spectrum when glycine is condensed [9], while another is the high binding energy of the $N(1s)$ electrons in the NH_3^+-group, as compared with in the neutral amino group [13]. The results of one previously published calculation of the atomic charge densities on the atoms of both neutral and zwitterionic glycine are shown in table 2. Note the large difference on the

Table 1. $\mathrm{H_3N^+C_\alpha H_2 COO^-}$ binding energies[a])

Sample	$N(1s)$	$C_\alpha(1s)$	$C(1s)$	$O(1s)$
Bulk glycine				
Clark et al. [4]	402.0	286.8	288.7	532.4
Present	402.2	286.7	288.7	532.3
Adsorbate on HOPG	402.5	286.7	288.9	532.5

[a]) Binding energies ± 0.2 eV

N-atom, which, in XPS, leads to the rather large binding energy of the $N(1s)$ electrons in the NH_3^+-group (at about 402 eV, see table 1), which in turn can be used as a measure of the presence of a zwitterionic form of glycine, figure 5.

The two carbon atoms in glycine also have unique binding energies relative to "neutral" carbon in, e.g., graphite, because of the net atomic charge densities in zwitterionic glycine, as seen in table 2. The presence of the final thin layer of glycine molecules on the surface of HOPG results in two weak but distinct peaks in the $C(1s)$ spectrum, of approximately equal intensity, which appear on the high binding energy side of the strong $C(1s)$ peak from C-atoms in the HOPG substrate. The intensity of the XPS signal from two glycine C-atoms can be enhanced relative to that of the HOPG substrate (fig. 3) by employing the surface geometry illustrated in figure 2. Thus the unique binding energies and relative intensities of the two small peaks in the $C(1s)$ spectrum confirm the presence of zwitterionic glycine on the HOPG surface.

The final core-level of interest is $O(1s)$. The clean HOPG substrate was essentially void of $O(1s)$ signal, whereas in the presence of glycine an $O(1s)$ peak is observed that is consistent, in binding energy and relative intensity, with the presence of zwitterionic

Table 2. CNDO/S Atomic charge densities of neutral and zwitterionic glycine [9]

Atom	Neutral	Zwitterion
H1	0.204	0.213
H2	0.140	0.228
H3	0.131	0.203
H4	0.055	0.057
C_α	0.069	0.021
C	0.417	0.395
N1	-0.352	0.044
O1	-0.452	-0.594
O2	-0.267	-0.624

glycine on the surface. The $O(1s)$ signal intensity, relative to that of the $N(1s)$ intensity, is not constant, however, as a function of θ. Qualitatively, the effect is similar to that seen in figure 5 in the $N(1s)$ spectra: the intensity of signal from the outermost atoms is enhanced as $\theta \to \theta_s$. In our XPS(θ) data, at $\theta = \theta_B$ the nitrogen-to-oxygen intensity ratio is very similar to that we measure for thick glycine films, whereas at $\theta = \theta_s$ the ratio is only about 0.7 of that at $\theta = \theta_B$. In other words, our data indicate that the molecules in the remaining glycine layer on HOPG tend to be preferentially oriented, with the O-atoms farther way from the surface and the N-atoms closer in to the surface. The intensity ratio of the $C(1s)$-to-$N(1s)$ shows a similar but weaker effect, and there is more scatter in the sample-to-sample data.

The total $C(1s)$ spectra, from the substrate and the over-layer, can be used in a more quantitative model, following Clark and coworkers, who studied the attenuation of XPS spectra of metal substrates with thin hydrocarbon polymer over-layers [6]. We make two simplifying assumptions: (a) the $C(1s)$ signal intensity from the HOPG substrate is attenuated by a homogeneous over-layer of thickness d; and (b) the attenuation in the thickness d is characterized by an energy-dependent mean-free-path for the escape of photoelectrons (the electron elastic-mean-free-path), λ_e, equal to that typical of hydrocarbon polymers. For electrons photoemitted from $C(1s)$ levels using Al k_α X-rays (1487 eV), the kinetic energy is about 1200 eV and $\lambda_e \approx 20$ Å [14]. Our XPS(θ) data are consistent with $d \approx 6(\pm 2)$ Å, where the standard deviation represents uncertainties due to counting statistics, baseline determinations, and sample-to-sample reproducibility.

Glycine crystallizes out of ethanol in the so-called β-form. The structure of single crystals of β-glycine has been determined by Iitaka [15]. The crystal structure consists of hydrogen-bonded double molecular layers, or sheets, with a unit cell dimension of 6.268 Å in the b-direction, that is, in the direction perpendicular to the hydrogen-bonded double molecular layers. The layers are extended in the two perpendicular directions. Successive double-layers are bonded only by weak van der Waals forces [15].

Our photoelectron spectra are consistent with the following picture of glycine on HOPG. A microcrystalline thick film of glycine condenses upon the HOPG surface as the ethanol evaporates in the vacuum system. Subsequently, excess glycine evaporates slowly, since the molecular over-layers are bonded to succeeding glycine layers only by weak van der Waals forces. Finally, a molecular double layer

similar, but no necessarily identical to, the $\beta\gamma$-glycine double layer remains upon the HOPG surface. The remaining layer can be removed by heating at 100°C, indicating that no real covalently bonded molecule-surface interaction has taken place. We postulate that the net (remaining) dipole moment of glycine in the double-layer is responsible for the surface adherence. The double-layer in β-glycine does have oxygen atoms preferentially toward the outside and nitrogen atoms preferentially toward the inside, but not to the extent indicated by our XPS(θ) data. Thus, the postulated double-layer of remaining glycine has a tendency for orientation in the direction normal to the surface, O-atoms out and N-atoms in toward the surface, but no lateral ordering is proposed.

The crystal structure of β-glycine is deliberately *not* shown here so as not to draw too close a parallel with glycine on HOPG. Indeed, the valence spectrum of the remaining glycine layers, at $\theta = \theta_s$ in figure 4, indicates some subtle electronic structural differences as compared with the valence band spectrum of bulk glycine. In addition, there are subtle but reproducible differences in the precise value of the core-level binding energies, $N(1s)$ in particular, in the thin remaining double-layer as compared with bulk glycine, see table 1. Simple electrostatic considerations indicate that an isolated glycine molecule should want to lie flat on the HOPG surface to maximize the induced electrostatic dipole interaction. "Flat" means that the $NC_\alpha COO$ backbone would be in the plane of the HOPG surface. The double layers observed in our study, however, do not have this orientation. This result indicates the importance of intra-layer intermolecular interactions, which should weaken the electrostatic interactions with the surface. For glycine condensed upon the HOPG surface at sub-monolayer coverage, the "flat" orientation might be expected.

Finally, we have begun an FTIR study of glycine adsorbed upon a variety of materials. These polarized-beam, reflection-absorption FTIR studies are carried out only under mild vacuum conditions, however, and thus the evaporation of glycine over-layers cannot be used to thin the samples. Thus far, solution-cast films of glycine down to 20 Å have been prepared. Our initial results can be analyzed using the well-known surface dipole selection rule [16]. Vibrational excitations polarized parallel to a metal surface are screened and are essentially not observed, whereas dipole excitations polarized perpendicular to the surface are observed strongly. The variations of the strength of the several NH_3^+ and COO^- modes of glycine on gold, relative to those for bulk glycine samples, indicate a preference for molecular orientation consistent with our proposed geometry [17, 18].

Conclusion

Starting with a motivation based upon biological considerations of protein adsorption onto non-biological surfaces, we have studied the adsorption of glycine, the smallest amino acid, upon a variety of non-biological surfaces. The most complete data set thus far is obtained for graphite or rather HOPG. On HOPG, when glycine is deposited from an ethanol solution, glycine molecules evaporate from the surface in a high vacuum environment until only approximately a double-layer of zwitterionic molecules remain. This remaining layer adheres to the HOPG surface in such a way that there is a tendency for molecular orientation in the direction normal to the surface, with the N-atom end of the molecule closer to the surface and the O-atom end away from the surface. There are subtle electronic differences between the double layer on HOPG and bulk glycine. The existence of the stable double-layer is consistent with the crystal structure of β-glycine, which is prepared by crystallization from ethanol. Polarized FTIR studies confirm the proposed geometry.

Acknowledgements

We acknowledge useful discussions with S. Welin. This project was partially funded by the Swedish Board for Technical Development (STU) through the Program in Surface Physics and Chemistry.

References

1. Lundström I, Salaneck WR (in press) J Colloid Interf Sci
2. Lundström I (1983) Physica Scripta T4:5
3. Wertheim GK, van Attekum PMTh, Basu G (1980) Solid State Commun 33:1127
4. Clark DT, Peeling J, Colling L (1976) Biochem Biophys Acta 453:533
5. Fadley CS, Baird RJ, Siekhaus W, Novakov T, Biegström SAA (1974) J Elec Spec 4:93
6. Clark DT, Dilks A, Shuttleworth D (1978) J Elec Spec 14:247
7. Thomas HR, O'Malley JJ (1979) Macromolecules 12:323
8. Debies TP, Rabalais JW (1974) J Elec Spec 3:315
9. Bigelow RW, Salaneck WR (1982) Chem Phys Lett 89:430
10. Biemann K, McCloskey JA (1962) J Am Chem Soc 84:3192
11. Kitaigorodsky AI (1973) Molecular Crystals and Molecules. Academic Press, New York
12. Aaron MW, Grant EH (1963) Trans Faraday Soc 59:85
13. Nordberg R, Albridge RG, Bergmark T, Eriksson U, Hedman J, Nordling C, Siegbahn K, Lindberg BJ (1967) Arkiv Kem 28:257
14. Clark DT (1981) In: Dwight DW, Fabish TJ, Thomas HR (eds) Photon, Electron and Ion Probes of Polymer Structure and Properties. Am Chem Soc, Washington, Chap 17, and references therein
15. Iitaka Y (1960) Acta Cryst 13:35
16. Greenler RG (1982) In: Gilles JM, Lucas AA (eds) Vibrations at Surfaces. Plenum Press, New York, p 255 – 264
17. Liedberg B, Ivarsson B, Lundström I, Salaneck WR (1985) Progr Colloid and Polym Sci, Vol 70
18. Liedberg B, Lundström I, Wu CR, Salaneck WR (to be published)

Received June 28, 1984;
accepted November 20, 1984

Authors' address:

W. R. Salaneck
Laboratory of Surface Physics
and Chemistry
University of Linköping
S-58183 Linköping (Sweden)

Progress in Colloid & Polymer Science   Progr Colloid & Polymer Sci 70:89 – 91 (1985)

Adsorption equilibria of surfactants on activated carbon from aqueous solutions

E. H. Stenby and K. S. Birdi

Fysisk-Kemiskt Institut, The Technical University of Denmark, Lyngby (Denmark)

Abstract: Studies on adsorption isotherms for anionic surfactant, n-$C_{12}H_{25}SO_4Na$, in aqueous solutions on activated carbon were carried out at different temperatures. The modified Freundlich adsorption theory was used to estimate the potential of adsorption. The isosteric heat of adsorption was calculated.

Key words: Adsorption, surfactants, solids, carbon, Freundlich

Introduction

The adsorption equilibria of surfactants on solid materials (e.g. minerals, carbon, etc.) is of much current interest, especially with regard to the surfactant-enhanced oil recovery phenomena. Even though a great many reports are found in the literature, much remains to be investigated due to the complexties involved at such interfacial phenomena. In general, an empirical relationship as proposed by Freundlich, equation 1, has been used by the majority of workers. A detailed description of various adsorption theories can be found in different text-books [1 – 3]. This study reports the adsorption equilibria of n-$C_{12}H_{25}SO_4Na$ (NaDDS) from aqueous solutions on carbon, as carried out at varying temperatures. The data were analyzed with the help of the modified Freundlich theory.

Theoretical

The well known empirical Freundlich equation which describes the adsorption process is given as [1 – 3]:

$$Q = kC_e^{1/n} \tag{1}$$

where Q is the molar (or weight) adsorbed amount per unit weight of adsorbent, C_e is the concentration of adsorbate in solution at equilibrium, k and n are the experimental parameters which depend on the system of adsorbent and adsorbate. In the case of activated carbon, the concentrations of adsorbate are very small, such that concentration units can safely be used instead of molar fractions [1 – 3].

The modified Freundlich equation gives the relationship between Q and C_e as follows [4]:

$$Q = k'(C_e/C_s)^{1/n} . \tag{2}$$

Where k' is the limiting amount adsorbed at the saturation concentration C_s (at monolayer coverage). The modified Freundlich equation and the expression for the adsorption potential, E_{eq}, was given as follows [4]:

$$Q = k' \exp(-E_{eq}/nRT) \tag{3}$$

where $E_{eq} = -RT\ln(C_e/C_s)$ with the following relation:

$$k' = kC_s^{1/n} . \tag{4}$$

From these equations one can then estimate the molar adsorption potential, E_{eq}, and its distribution function. The isosteric differential heat of adsorption, q_{iso}, can be derived from the Clausius-Clapeyron equation and the modified Freundlich equation:

$$q_{iso} = RT^2(d\ln C_e/dT)_Q \tag{5}$$

$$= -nRT^2(d\ln k'/dT) - RT^2(dn/dT)\ln k'$$
$$+ RT^2(dn/dT)\ln Q + RT^2(d\ln C_s/dT) . \tag{6}$$

If k' (mol/g) is temperature independent, then the relation in equation (5) reduces to [4]:

$$q_{\text{iso}} = E_{\text{eq}} + RT^2(d\ln C_s/dT) . \qquad (7)$$

Under the assumption that the heat of dissolution of adsorbate into solution, q_{diss}, is equal to $-RT^2(d\ln C_s/dT)$, then the isosteric differential heat of adsorption can be estimated from equation (8):

$$q_{\text{iso}} = E_{\text{eq}} - q_{\text{diss}} . \qquad (8)$$

In other words, the isosteric heat is equivalent to the adsorption potential, E_{eq}, minus the heat of dissolution of adsorbate into solution.

Experimental and procedure

The adsorption isotherms were studied by a batch adsorption method. To 100 ml of water was added 0.1 – 0.5 g of activated carbon with vigorous mixing. The surfactant solution (50 g/L) (n-$C_{12}H_{25}SO_4Na$, NaDDS, as supplied by Serva, W. Germany, purity >99%) was added in 100 μl aliquats. The concentration of surfactant was monitored by using a conductivity meter (Radiometer, Denmark). Equilibrium was observed to be reached after 15 minutes of each addition, as found from the conductivity. The temperature was kept constant. The concentration of free NaDDS was determined from conductivity measurements.

Results and discussion

The adsorption isotherms for NaDDS on carbon are shown in figure 1, i.e. $\log Q$ versus $\log C_e$, equation (2). These plots are linear, with the exception of some departure from linearity in the lower concentration region. Similar behavior has been reported by other investigators when studying the adsorption of organic molecules on carbon, such as benzene, phenol, aniline, etc. [4].

The magnitude of k' was 11.9 (± 0.3) 10^{-4} mol/g and of $n = 2.5 \pm 0.5$, in the temperature range 22 – 39 °C. The values of E_{eq} and q_{iso} as a function of temperature are given in figure 2. In the case of surfactant systems as used here, the quantity $C_s = \text{CMC}$ (critical micelle concentration) and thus equation (7) can be rewritten as:

$$q_{\text{iso}} = E_{\text{eq}} + RT^2\left(\frac{d\ln \text{CMC}}{dT}\right) . \qquad (7')$$

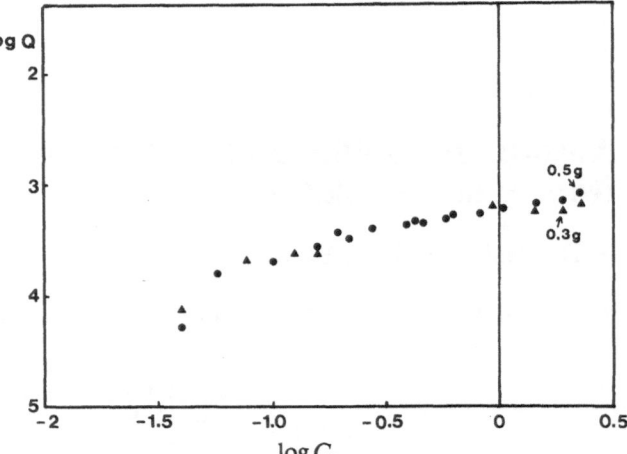

Fig. 1a

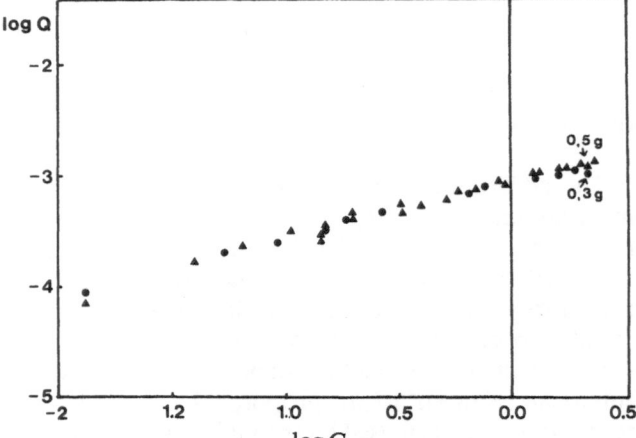

Fig. 1b

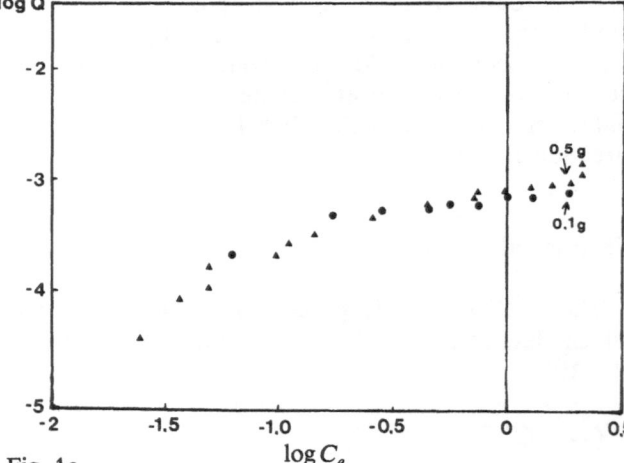

Fig. 1c

Fig. 1. Variation of $\log Q$ with $\log C_e$ for the system: NaDDS-carbon (a: 22 °C, b: 33 °C, c: 39 °C). Amount of carbon is indicated in the figure.

$$\left[Q = \frac{\text{moles NaDDS}}{\text{g solid}} ; C_e = \text{g/L} \right]$$

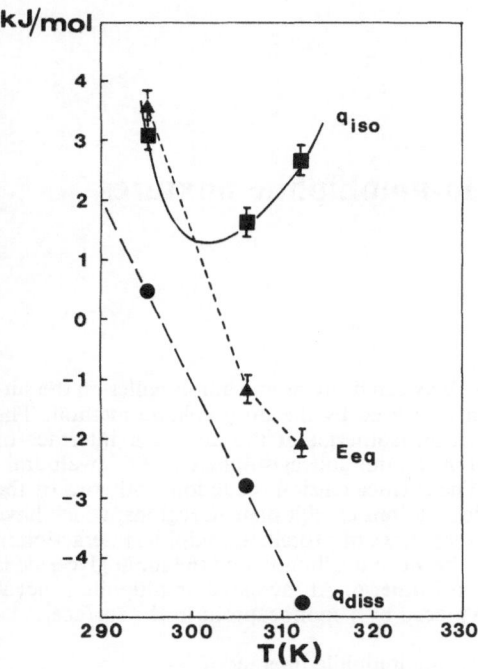

Fig. 2. Variation of q_{iso} (■) and E_{eq} (▲) and
$$\left[-RT^2 \left(\frac{d \ln CMC}{dT} \right) \right]$$ (●) with temperature for the
system: NaDDS-carbon

It is known that the right-hand-term in equation (7′)
varies linearly with temperature for NaDDS solutions
around 25 °C [5], figure 2.

The magnitude of E_{eq} has been found to be related
to the interaction energy between the adsorbant and
the adsorbate, i.e., E_{eq} for phenol was larger than that
of benzaldehyde on carbon [4]. This observation was
explained as arising from the possibility of specific interactions between active sites on carbon and phenol
[4]. Further studies are in progress in order to determine the significance of E_{eq} and q_{iso}.

Acknowledgements

E. H. Stenby would like to thank the Danish Council for
Scientific and Industrial Research (STVF) for the research
grant (licentiatstipendium).

References

1. Kipling JJ (1965) Adsorption from Solutions of Non-
   Electrolytes. Academic Press, New York
2. Adamson AW (1976) Physical Chemistry of Surfaces, 3rd
   ed. Academic Press, New York
3. Chattoraj DK, Birdi KS (1984) Adsorption and the Gibbs
   Surface Excess. Plenum Press, New York
4. Urano K, Koichi Y, Nakazawa Y (1981) J Colloid Interface Sci 81:477
5. Mysels KJ, Mukerjee P (1971) Critical Micelle Concentrations of Surfactant Systems. NSRDS, NBS-26, USA,
   Feb 1971. Washington, DC

Received July 2, 1984;
accepted November 5, 1984

Authors' address:

K. S. Birdi
Fysisk-Kemiskt Institut
Technical University of Denmark
Buildn. 206
DK-2800 Lyngby (Denmark)

Progress in Colloid & Polymer Science   Progr Colloid & Polymer Sci 70:92 – 95 (1985)

Surface behaviour of adsorbed films from protein-amphiphile mixtures

B. Ericsson and P. O. Hegg

Dept. of Food Technology, University of Lund, Lund (Sweden)

Abstract: The effect of interaction between proteins and amphiphiles on the surface tension reduction have been measured by the drop-volume method. The equilibrium surface tension reduction isotherms at the air/water interface of serum albumin and of ovalbumin in sodium dodecylsulphate and of ovalbumin in 1-monocaproin are reported. The surface tension reduction isotherms of the proteins in the anionic amphiphile solutions exhibit plateau regions, which have been interpreted in terms of different states of protein-amphiphile interaction in the bulk solution. Any interaction between ovalbumin and the monoglyceride is not reflected in the surface tension isotherm. At increased amphiphile concentration the protein seems to be replaced by 1-monocaproin in the surface.

Key words: Surface tension, protein-amphiphile interaction

Introduction

Interaction between proteins and other amphiphilic substances has been shown to influence technical properties of proteins in food systems. Since proteins and low molecular weight amphiphiles often coexist in systems where formation and properties of interfacial films are of fundamental importance (e.g. dairy products, dough, meat emulsions), influence of the mode of interaction between proteins and amphiphiles in the bulk solution on the surface behaviour is of interest.

The interfacial tension of adsorbed protein films in connection with functional properties has been reported [1, 2, 3]. Also the interfacial mixed films of proteins and lipids have been investigated, either by spreading films of both components [4, 5], or by examination of the penetration of one component into a surface film of the other component [6, 7].

In the present study, the equilibrium surface tension reduction of adsorbed films from mixtures of proteins and lipids possessing different interactions in bulk solution are reported. The systems investigated are bovine serum albumin (BSA) and ovalbumin (OA) in sodium dodecylsulphate (SDS) and OA in 1-monocaproin. The interaction between SDS and proteins are well documented in literature and reviewed in reference [8]. SDS is known to bind to most proteins in a nonspecific, co-operative manner, thereby gen-

erally inducing gross unfolding of the protein structure. Few proteins seem to exhibit specific binding sites for a lipid type of amphiphiles. Studies on the interaction between BSA and SDS have revealed 8 – 10 high affinity sites, which are both hydrophobic and electrostatic in nature [9]. The specific binding of anionic amphiphiles to BSA stabilizes the protein structure, which can be seen as an increased resistance towards thermal denaturation [10]. At SDS-to-BSA ratios above the specific binding region the interaction is nonspecific, resulting in unfolding of the protein structure. OA, on the other hand, interacts with SDS only in a nonspecific mode [11].

In the third system studied, OA in 1-monocaproin solution, none or moderate interaction is supposed to occur, since it has been shown [12, 13] that non-ionic amphiphiles only possess weak affinity, if any, to water-soluble proteins. 1-Monocaproin has interesting phase behaviour in aqueous systems. Only one liquid phase exists [14]; in water-rich systems ordinary micelles are formed and in amphiphile-rich systems the isotropic liquid phase is of the L2 type. The interfacial behaviour of this component has not yet been reported.

Materials and methods

Bovine serum albumin (essentially fatty acid free, lot nr. 16C-7201) and hen egg white ovalbumin, free from the S-variant [15] (lot nr. 116C-8220) were purchased from Sigma

Chemical Co. The specifications given by the manufacturer
of the proteins were checked by differential scanning calo-
rimetry. Both proteins gave single transition peaks revealing
no fatty acids in the BSA preparation [10] and no con-
tamination of the S-variant in the ovalbumin preparation
[16]. SDS-polyacrylamide electrophoresis showed single
bands for both proteins. BSA was used as supplied, while
ovalbumin was desalted on a column of Sephadex G-25 (medium) in ammonium bicarbonate buffer. After chromatography and dialysis against diluted buffer, the protein was
freeze-dried to constant weight.

Sodium dodecyl sulphate (SDS), specially pure (declared
purity 99%) was from BDH Chemicals Ltd., and used as
supplied. The preparation of 1-monocaproin has been described by Larsson [17]. Other reagents used were of analytical grade.

Protein solutions, made up in either phosphate buffer or
in double distilled water, was diluted to final concentration
(1 mg/ml) by addition of amphiphile solution of appropriate concentration. Protein and monocaproin solutions were freshly prepared, while the diluted SDS solutions
were made up from a stock solution (0.5 mM). To avoid precipitation at low molar ratios of the anionic amphiphile-pro-
tein complexes, the experiments were performed at pH 5.6,
i.e. at the alkaline side of the isoelectric point of the proteins. All measurements were performed at 23 ± 0.5 °C.

The time dependent surface tension decay was measured
according to the drop-volume technique as outlined by
Tornberg [18, 19]. The automatization procedure according
to Arnebrant and Nylander [20] was employed. In this method surface tension reduction by macromolecules during adsorption at the air-water interface is measured by formation
of drops of certain volumes. Time for detachment of the
droplets is recorded. Surface tension calculated [19, 20] was
plotted against detachment time and the value attained after
2000 seconds was set as the equilibrium value. The surface
tension of the solutions is still decreasing after this period of
time, but the rate of decrease is small, less than 0.05 mNm^{-1}
per 100 seconds. The maximum error in surface tension
values is ± 1.5%.

The drop-volume method allows measurements of solutions without further dilution, i.e. irreversibility and/or
time-dependent reversibility at dilution of the complexes due
to total concentration of amphiphile is avoided. In addition,
adsorption to the surface created during formation of the
droplet might be seen as analogous to adsorption to surfaces
created in foams and emulsions, for example.

Results and discussion

The surface tension reduction ($\Delta\gamma$) by SDS of
0.05 M phosphate buffer, pH 5.6, at 23 °C is
indicated in figures 1 and 2. At conditions used surface tension of SDS at concentrations above 1 mM is
not possible to measure, since at this concentration
the critical temperature curve is surpassed. The knee-
point of this curve, the Krafft point, which has been
reported to 9 – 10 °C [21], is known to be highly sensitive to impurities, which decrease the critical temperature, and to added salts, which is thought to increase
this temperature [22].

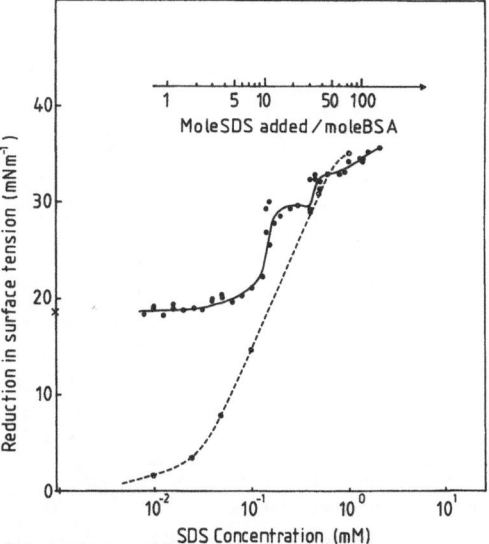

Fig. 1. The equilibrium surface tension reduction isotherm
of SDS (broken line) and of 13.4 µM BSA in SDS solution
(solid line). The surface tension reduction of 13.4 µM BSA
without SDS is indicated on the ordinate. Measurements
were performed in 0.05 M phosphate buffer, pH 5.6 at 23 °C

The equilibrium value of reduction in surface tension by 13.4 µM BSA of the phoshate buffer is
19.4 mNm^{-1}. The protein concentration is well above

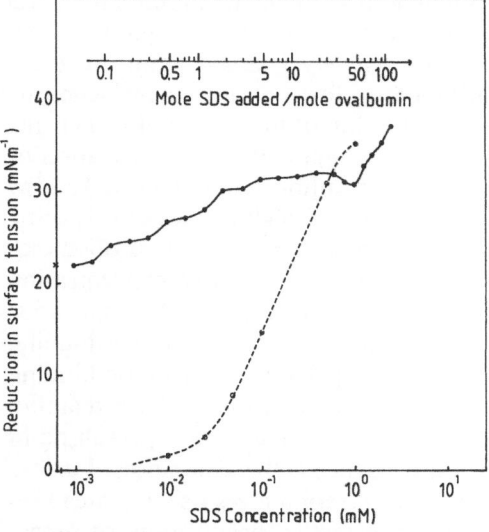

Fig. 2. The equilibrium surface tension reduction isotherm
of SDS (broken line) and of 21 µM ovalbumin in SDS solution (solid line). The surface tension reduction of 21 µM ovalbumin without SDS is indicated on the ordinate. Experimental conditions as in figure 1

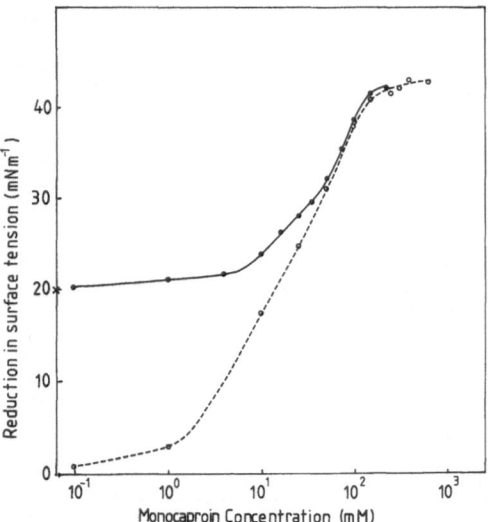

Fig. 3. The equilibrium surface tension reduction isotherm of 1-monocaproin (broken line) and of 21 μM ovalbumin in 1-monocaproin solution (solid line). The surface tension reduction of ovalbumin without the amphiphile is indicated on the ordinate. Measurements were performed at 23 °C in double distilled water, pH adjusted to 5.6 using 1.0 N NaOH

the surface saturation concentration. $\Delta\gamma$ of 13.4 μM BSA as a function of total SDS concentration is shown in figure 1. At SDS-to-BSA ratios approximately corresponding to the specific binding region, no significant effect on surface tension can be seen. The specific binding of anionic amphiphiles to serum albumin is supposed to occur to hydrophobic areas situated close to positively charged amino acid residues [9]. This binding does not seem to influence the saturation surface tension of the protein. Further, the association constant is high and the concentration of unbound SDS in the solution is low (below 10^{-5} M [8]). At SDS concentrations slightly above the specific binding region, an abrupt increase in $\Delta\gamma$ is observed. This might be caused by the increased concentration of unbound SDS. Furthermore, at SDS concentrations below those required for co-operative binding (approximately 0.1 mM [13]) an electrostatic binding of the anionic amphiphile to positively charged amino acid residues of the protein might occur, resulting in formation of more surface active complexes. Further increase of SDS concentration gives rise to plateau regions, indicating constant concentration of unassociated amphiphile. This behaviour is to be expected at the co-operative association of amphiphile to protein, occurring over a very narrow range of free SDS concentration. This binding has also been shown to be

accompanied by an unfolding of the protein structure [8, 10]. But still both unbound SDS and the complex formed are supposed to contribute to the surface tension reduction, since the $\Delta\gamma$-value at the plateau is higher than the $\Delta\gamma$-value for pure SDS at equal concentrations.

At SDS concentrations above approximately 1 mM the surface tension reduction seems to be mostly governed by the amphiphile. Either no further interaction to the protein occurs, or the free SDS concentration is sufficiently high to create surface films that cannot be penetrated by the negatively charged protein-amphiphile complex.

The reduction in surface tension by 21 μM ovalbumin of the phosphate buffer is 22 mNm^{-1}, i.e. a slightly higher value than obtained by BSA. The $\Delta\gamma$ isotherm of 21 μM ovalbumin as a function of SDS concentration is shown in figure 2. Except for the region of specific binding in the BSA-SDS surface tension isotherm, the general pattern is similar. At low SDS concentration, where no co-operative binding of SDS to OA is supposed to occur, there is a gradual increase in $\Delta\gamma$. This phenomenon can be caused by a synergistic effect, arising from the possibility of a more optimal packing of SDS and protein segments at the surface compared to the individual components. Further, an electrostatic interaction between the protein and SDS would increasing the hydrophobicity of the protein. Most probably, the increase originates from a combination of both these processes. At increased SDS concentration, a plateau value of $\Delta\gamma$ is reached, where, in analogy to the BSA-SDS system, a co-operative binding of SDS to OA, associated with unfolding of the protein, is supposed to occur. For OA this region starts at a lower total SDS concentration, while the effect of interaction on the surface tension isotherm is completed at around the same total concentration of amphiphile.

Figure 3 shows the surface tension reduction isotherm of 1-monocaproin in double distilled water. The association concentration of this compound is high, the critical micelle concentration is estimated to 160 mM. The saturation value of $\Delta\gamma$, 42 mNm^{-1}, though, is in the same order of magnitude as obtained by long-chained monoglycerides. The presence of ovalbumin at a concentration of 21 μM only displaces the surface tension isotherm to the $\Delta\gamma$-value of the pure protein in water (20 mNm^{-1}). At amphiphile concentrations below 5 mM only a slight increase in $\Delta\gamma$ is seen, i.e. the free nonionic amphiphile does not seem to contribute to the surface tension reduction. This might be interpreted as an interaction between the protein and monocaproin not influencing the surface tension decrease. But since no interaction between a nonionic amphiphile and water-soluble proteins is supposed to occur [12, 13], it is more likely that the protein, being the most surface-active component at low concentration of this amphiphile, dominates the surface film. Further, no synergism in surface tension is reported for mixtures of nonionic amphiphiles [23]. The surface tension reduction at higher monocaproin concentrations seems to be mostly governed by the amphiphile, the protein merely being replaced in the surface film by the monocaproin. At surface tension reduction above 30 mNm^{-1} the isotherm in presence of protein is identical to the one seen by pure monocaproin, indicating that all the protein molecules have been completely squeezed out from the surface film. This can also be seen as a support for the proposed lack of interaction between the protein and amphiphile since there is no displacement of the isotherm towards higher amphiphile concentration in presence of protein. Such a displacement is clearly seen in the other two isotherms presented.

Conclusions

From the surface tension isotherms reported it can be generalized that interactions between protein and amphiphile in the bulk solution are closely related to adsorption behaviour at the air/water interface. Association of lipid-like substances to proteins results in plateau regions of the isotherms, within which neither changes in the amphiphile concentration, nor the supposed consecutive unfolding of the protein structure is reflected in the $\Delta\gamma$-value. Replacement of protein in the surface film by a lipid type of amphiphile, when no interaction occurs, was seen in the monocaproin-ovalbumin isotherm, and also in the SDS-protein isotherms at sufficiently high amphiphile concentrations. Similar effects have also been observed in the membrane of the fat globule in milk [24], by addition of a nonionic amphiphile at emulsification.

References

1. Graham DE, Phillips MC (1979) J Colloid Interface Sci 70:403, 415, 427
2. Phillips MC (1981) Food Technol 35:50
3. Tornberg E (1978) J Sci Fd Agric 29:762
4. Tabak SA, Notter RH (1977) J Colloid Interface Sci 59:293
5. Yudenfreund MV, Becher P (1975) Adv Chem Ser 144:192
6. Colocicco G (1969) J Colloid Interface Sci 29:345
7. Wiedmer T, Brodbeck U, Zahler P, Fulpius BW (1978) Biochim Biophys Acta 506:161
8. Steinhardt J, Reynolds J (1969) Multiple Equilibria in Proteins. Academic Press, New York
9. Tanford C (1972) J Mol Biol 67:59
10. Gumpen S, Hegg PO, Martens H (1979) Biochim Biophys Acta 574:189
11. Yang JT, Foster JF (1953) J Am Chem Soc 75:5560
12. Clarke S (1975) J Biol Chem 250:5459
13. Makino S, Reynolds JA, Tanford C (1973) J Biol Chem 248:4926
14. Larsson K (1967) Z Phys Chem 56:173
15. Nakamura R, Hirai M, Takemori Y (1980) Agric Biol Chem 44:149
16. Hegg PO, Martens H, Löfqvist B (1979) J Sci Fd Agric 30:981
17. Larsson K (1964) Ark Kemi 23:35
18. Tornberg E (1977) J Colloid Interface Sci 60:50
19. Tornberg E (1978) J Colloid Interface Sci 64:391
20. Arnebrant T, Nylander T (in press) J Disp Sci Technol
21. Hato M, Shinoda K (1973) J Phys Chem 77:378
22. Fontell K (1981) Mol Cryst Liq Cryst 63:59
23. Garrett PR (1975) J Chem Soc Faraday Trans I 72:1327
24. Oortwijn H, Walstra P, Mulder H (1977) Neth Milk Dairy J 31:134

Received July 13, 1984;
accepted October 15, 1984

Authors' address:

Bodil Ericsson
Dept. of Food Technology
Chemical Center
University of Lund
Box 124
S-22100 Lund (Sweden)

Biosensors based on surface concentration measuring devices –
The concept of surface concentration

U. Jönsson[1], M. Malmqvist[2], I. Rönnberg[3], and L. Berghem[3]

[1]Laboratory of Applied Physics, [2]Division of Radiation Biology, [3]Division of Experimental Medicine,
National Defence Research Institute, Department 4, Umeå (Sweden)

Abstract: Surface concentration measuring devices allow direct measurement of biomolecular interactions. These devices measure the surface concentration of, for example, interacting antigens or antibodies at solid surfaces.

We here exemplify the importance of using highly purified molecules together with an optimized immobilization technique for surface concentration measuring immunosensors. The immobilization of antibodies for the detection of human plasma fibronectin was compared to the reverse where the antigen was immobilized. The result was evaluated by *in situ* ellipsometry, an optical surface concentration measuring technique whereby the changes in polarization of light on reflection at a surface is measured.

It was found that immobilized fibronectin to a much greater extent interacted with antibodies in solution as compared to the reverse situation where the antibodies were immobilized. This was explained by the multivalency of antibody binding sites on each fibronectin molecule as compared to the two antigen binding sites on each antibody. The surface concentration of fibronectin interacting with immobilized antibodies was found to depend not only on the degree of purification of the antibodies but also on the immobilization technique.

Key words: Biosensor, surface concentration, ellipsometry, surface test for antigen, antibody-immobilization

Introduction

New surface concentration measuring biosensors, based on an affinity reaction between enzyme/substrat or antibody/antigen for example, focus the interest on surface modification, immobilization techniques and fundamental knowledge about biomolecular interactions at solid surfaces. Various surface concentration measuring devices [1 – 6] can be used to monitor either the kinetics or the equilibrium of interaction without the introduction of radioactive, fluorescent or enzyme labels. The devices are independent of the size of the solid surface thus allowing miniaturization and the use of small sample volume. This property, however, requires that a maximum number of immobilized molecules are active. The immobilization should therefore be optimized avoiding denaturation but also allowing a proper orientation of the immobilized molecules. In conventional immunological solid phase assays such as radio-immunoassay [7] or enzyme linked immunosorbent assay [8] the total amount of interacting

molecules is measured. The sensitivity in these assays is therefore dependent on the size of the solid phase.

We here exemplify the importance of an optimized immobilization technique for surface concentration measuring immunosensors by using human fibronectin (HFN) and rabbit antibodies. The result was evaluated by *in situ* ellipsometry, an optical surface concentration measuring technique whereby the changes in polarization of light on reflection at a surface are measured.

HFN is associated with the cell surface of most fibroblast and endothelial cell lines and is also a normal circulating plasma protein (for review see Yamada [9]; Saba and Jaffe [10]).

Several biological activities have been proposed for HFN. These activities include the ability to mediate cell-substratum and cell-cell interaction, to promote wound healing and to function as an opsonin. The study of *in vivo* roles for HFN under normal and pathological conditions requires rapid, simple and reproducible methods for obtaining HFN levels in plasma and other biological fluids.

Materials and methods

Single crystalline silicon (Wacker Chemitronic, FRG) was cut in 3×3 mm slides and washed according to Jönsson et al. [11]. The washed slides were blown dry in nitrogen gas and silanized by a modified version of the chemical vapour deposition technique described in [12] of (3-((2-Aminoethyl)-amino)-propyl)trimethoxysilane (Union Carbide, USA). The main difference was that the slides to be silanized and the vacuum chamber could be held at a desired elevated temperature. Details of this silanization technique may be found in [13]. Pyridyldisulfide groups were introduced to these surfaces by reaction with 5 mM of the reagent SPDP (Pharmacia Fine Chemicals, Sweden) in 0.1 M sodium phosphate, 0.1 M NaCl and 1 mM EDTA, pH = 7.5, for 30 minutes. This buffer was used throughout in all incubations. These slides are in the following denoted "pyridyl slides". Such slides were rinsed and reduced with dithiothreitol (Aldrich, Belgium), 100 mM, for 30 minutes. After rinsing, the slides were incubated for 12 hours with SPDP-modified Protein A (Pharmacia Fine Chemicals, Sweden) according to Carlsson et al. [14] at a concentration of 100 µg/ml. The degree of SPDP-modification was on average 2.5 SPDP/Protein A molecule.

To avoid transfer of impurities from the air-liquid interface to the surfaces and also to ensure a proper rinsing after the various incubation steps, all handling of the slides took place with a continuous flow of buffer. This was accomplished by continuous back flow of buffer through a glass filter funnel thereby causing an overflow of buffer from the funnel.

HFN and rabbit antibodies to HFN were purified and characterized as described before [11]. The antibodies were either the immunoglobulin G fraction (IgG-Anti-HFN) or the specific antibodies obtained from immunoadsorption on HFN immobilized to Sepharose (Anti-HFN) [11]. Protein A was purchased from Pharmacia Fine Chemicals, Sweden. HFN, IgG-Anti-HFN, Anti-HFN or the immunoglobulin G fraction from non immunized rabbits (IgG) were adsorbed for 18 hours to pyridyl slides at a concentration of 100 µg/ml respectively. Slides with covalently immobilized Protein A were incubated with Anti-HFN or IgG, 100 µg/ml, for 2 hours. The antigen/antibody interactions were studied by incubations of the slides for 2 hours with added concentrations of 0, 3 and 10 µg/ml respectively of HFN, Anti-HFN or IgG.

The petri dishes together with all protein handling equipment were washed overnight in Hellmanex (Hellma, G.D.R.) followed by rinsing in water. All chemicals were of analytical grade, the water used was double distilled and the reactions were made at room temperature. The protein incubations were made in polystyrene petri dishes as described in [18]. Inherent in this technique is the difficulty of clearly defining the true protein concentration in solution due to unspecific adsorption to container walls and the air-liquid interface. The area/volume ratio for a dish including the air/liquid interface with no silicon slides was roughly 0.63 mm^2/mm^3. This ratio increases with the silicon slides to 0.65 indicating that a large surface area is available for unspecific adsorption. If the unspecific adsorption is assumed to be of the order 5 ng/mm^2 the protein concentration in solution below an added concentration of 10 µg/ml will be uncertain. As the surface to volume ratio is kept constant between different experiments in this investigation compari-

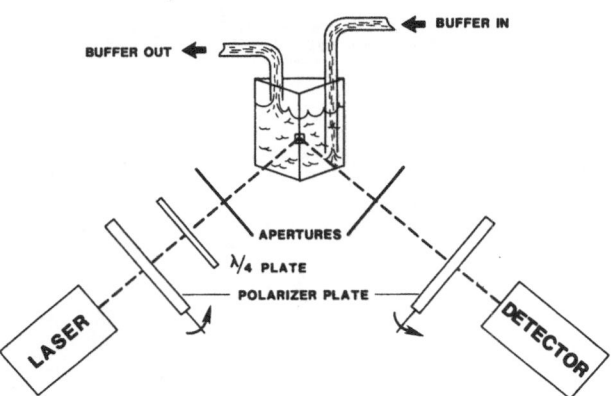

Fig. 1. *In situ* ellipsometry. A laser beam of light is reflected against the solid sample surface at an angle of incidence of 60 degrees. Before reaching the surface the light passes through a polarizer and a quarter of wave length plate. After reflection the light passes through a second polarizer and is detected by a photodiode. The polarizers are automatically rotated in a way that the resulting light intensity reaching the detector is kept at a minimum. The position of the polarizers gives the angles Ψ and Δ

son between the experiments can be made. It should, however, be kept in mind that the true protein concentration in solution is certainly smaller than the added concentration.

The ellipsometer (fig. 1) measures the change in the polarization of light upon reflection on a surface. From this change the optical constants of the surface can be determined. The technique has been used in numerous investigations (e.g. [15, 16]) for the study of protein/solid surface interactions. For further information about theory, instrumentation, and calculation see [17]. Ordinary null-ellipsometry gives two angular readings, which are converted into changes in amplitude, tan Ψ, and phase, Δ, of light upon reflection. If these two parameters are measured for a reflecting surface the optical constants of the surface can be determined. If a film of molecules is adsorbed onto the surface, tan Ψ and Δ will change and from the new values refractive index, layer thickness and amount of absorbed material may be calculated. Due to the optical properties of silicon it is mainly Δ that changes when proteins are adsorbed or interacting at a surface. Because of the reproducibility of this parameter for silicon [11] it was possible to first measure all slides in the ellipsometer, secondly put them in the various incubation media and finally after rinsing again measure them in the ellipsometer.

The exact amount of interacting or adsorbing proteins was in parallel experiments determined by simultaneous ellipsometric and radiometric measurements of ^{14}C-labelled HFN, IgG and Protein A. An empirical relationship between the ellipsometric parameter Δ and the amount of proteins on the surfaces could in this way be obtained [18].

The ellipsometer used was a Rudolph Research ellipsometer (Model 43603) modified for automatic null-ellipsometry. All measurements were performed without drying of the silicon slides.

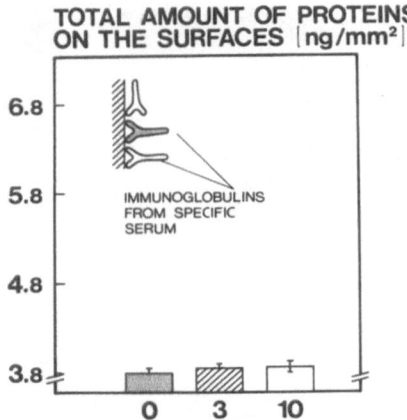

Fig. 2. The interaction of adsorbed IgG-Anti-HFN with HFN. The amount of proteins on the surface is presented as a function of 0, 3 and 10 µg/ml respectively of added HFN in solution

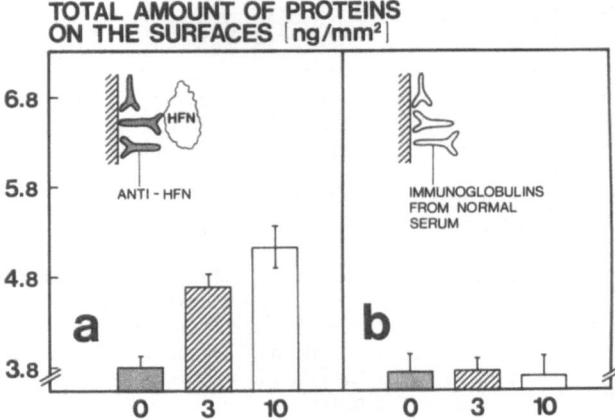

Fig. 3. The interaction of adsorbed Anti-HFN (a) and IgG (b) with HFN. Explanations, see figure 2

Results

The chemical vapour deposition of silane to the silicon surface minimizes the risk of silane polymerization [19, 20]. Our technique gives smooth silane layers of monolayer character as measured by ellipsometry. With an assumed refractive index for the silane of 1.46 [19] the thickness of the silane layer on the slides is measured by ellipsometry to be 6 ± 1 Å.

Figures 2 – 4 show the interaction of adsorbed IgG-Anti-HFN, Anti-HFN, and Anti-HFN immobilized to covalently surface attached Protein A with HFN at added concentrations of 3 and 10 µg/ml respectively.

The experiments show that significant amounts of HFN could be detected when antibodies purified by immunoadsorption were used (fig. 2 as compared to fig. 3a). The immobilization of such antibodies to Protein A increased the amounts of interacting HFN as compared to the adsorption of Anti-HFN to the bare surface (fig. 3a as compared to fig. 4a).

To ascertain that the changes in adsorbed amounts are not due to unspecific adsorption to the surfaces the experiments were repeated with immunoglobulins obtained from normal rabbit serum (figs. 3b and 4b). These experiments strongly suggest that the changes are due to the specific antibody/antigen interaction.

Figure 5a shows the interaction of Anti-HFN with a preadsorbed film of HFN. The specificity of the interaction was ascertained by experiments performed with IgG (fig. 5b).

The figures show one experimental run performed on three different slides. The bars represent the mean values from these slides. The maximum deviations form these mean values are also indicated. Minor quantitative differences $\pm 15\%$ exist between different runs. The qualitative behaviour was, however, the same.

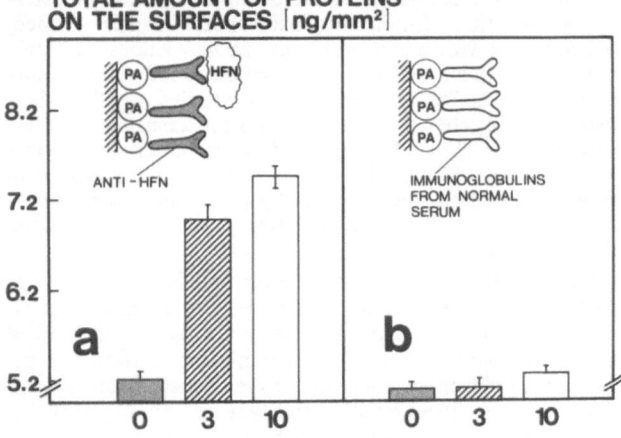

Fig. 4. The interaction of Anti-HFN (a) and IgG (b) immobilized to covalently surface attached Protein A with HFN. Explanations, see figure 2

Discussion

The study of immunological interactions by ellipsometry was pioneered by Rothen [21]. These studies were successful for the detection of antibodies via surface immobilized antigen. However, when the situation was reversed and the antibodies were immobilized to a surface it was not possible to detect protein antigens.

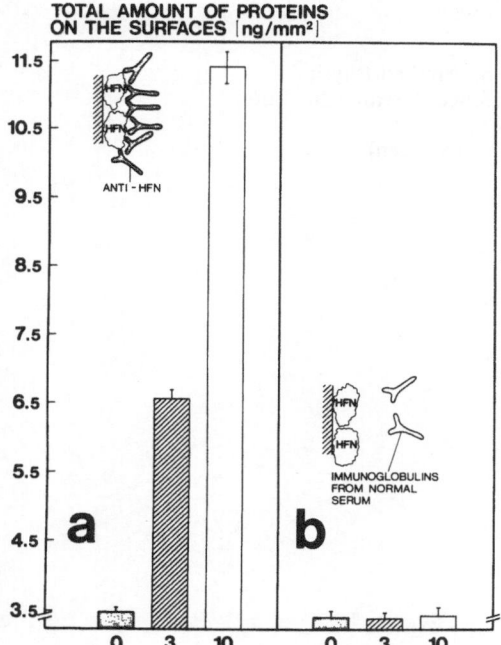

Fig. 5. The interaction of adsorbed HFN with Anti-HFN (a) and IgG (b). The amount of proteins on the surface is presented as a function of 0, 3 and 10 µg/ml respectively of added Anti-HFN (a) and IgG (b) in solution

The importance of surface concentration was discussed by Nygren et al. [5]. Their experiments indicated that direct optical visualization of a second protein layer applied upon a pre-existing one is possible only if the second layer reached a certain surface concentration that has a typical threshold value for each visualization system.

Our experimental system is based on HFN and antibodies from the immunoglobulin G fraction before and after affinity purification. The immunoglobulin G fraction from an immunized rabbit contains not only HFN specific antibodies but also antibodies to other antigens. The HFN specific antibodies can be purified by affinity chromatography. The surface concentration of HFN interacting with surface immobilized antibodies is thus dependent on the purity of the antibodies (figs. 2 and 3).

Another parameter expected to be of importance for surface concentration measuring devices is the orientation of active molecules on the surface (fig. 4). The Protein A is expected to specifically bind Anti-HFN in its non-antigen binding part [22]. The Protein A may thus orient the Anit-HFN in a manner that the antigen binding part of the molecule is directed away from the surface. The binding of Anti-HFN to Protein A is also expected to be a more gentle treatment

as compared to the adsorption where changes in conformation and/or partial denaturation may occur. In control experiments it was shown that the Anti-HFN bound to Protein A could be completely desorbed by lowering the pH to 2.5. Only about 15% of the surface-adsorbed Anti-HFN were desorbed by this treatment, strongly suggesting that the Anti-HFN are bound to Protein A and not to free adsorption sites on the Protein A covered surfaces.

High molecular weight protein antigens have in general several antibody binding sites, so-called epitopes. Even a randomly oriented immobilized antigen may thus expose one or several epitopes into the solution. On the other hand, antibodies obtained from the immunoglobulin G fraction possess two antigen binding sites. Furthermore these sites are located in one part of the antibody molecule.

The differences in surface concentration of interacting molecules depending on whether the antigen or the antibodies are surface immobilized (figs. 2 – 4 as compared to Fig. 5) may therefore be explained by the multivalency of antibody binding sites on each HFN molecule as compared to the two HFN binding sites on each specific antibody.

References

1. Giaever I (1976) J Immunol 116:766
2. Yamamoto N, Nagasawa Y, Shuto S, Tsubomura H, Sawai M, Okumura H (1980) Clin Chem 26:1569
3. Janata J, Huber RJ (1980) In: Freiser H (ed) Ion-Selective Electrodes in Analytical Chemistry, Vol 2. Plenum Press, New York, p 107
4. Nylander C, Liedberg B, Lind T (1982 – 83) Sensors and Actuators 3:79
5. Nygren H, Sandström T, Stenberg M (1983) J Immunol Methods 59:145
6. Suzuki S, Karube I (1981) Appl Biochem Bioengin 3:145
7. Catt K, Tregear GW (1967) Science 158:1570
8. Engvall E, Perlmann P (1971) Immunochemistry 8:871
9. Yamada KM (1980) Lymphokine Rep 1:231
10. Saba TM, Jaffe E (1980) Am J Med 68:577
11. Jönsson U, Ivarsson B, Lundström I, Berghem L (1982) J Colloid Interface Sci 90:148
12. Mittal KL, O'Kane DF (1976) J Adhes 8:93
13. Jönsson U, Malmqvist M, Olofsson G (1982) Swedish Patent Appl No 8200442-5
14. Carlsson J, Drevin H, Axén R (1978) Biochem J 173:723
15. Vroman L, Adams AL (1969) Surface Sci 16:438
16. Cuypers PA, Corsel JW, Janssen MP, Kop JMM, Hermens WTh, Hemker HC (1983) J Biol Chem 258:2426
17. Azzam RMA, Bashara NM (1977) Ellipsometry and Polarized Light. North Holland, Amsterdam
18. Jönsson U, Malmqvist M, Rönnberg I (1985) J Colloid Interface Sci
19. Haller I (1978) J Am Chem Soc 100:8050
20. Bascom WD (1972) Macromolecules 5:792

21. Rothen A (1945) Rev Sci Instrum 16:26
22. Gyka G, Ghetie V, Sjöqvist J (1983) J Immunol Methods 57:227

Received June 1, 1984;
accepted September 5, 1984

Authors' address:

Ulf Jönsson
Laboratory of Applied Physics
National Defence Research Institute
Department 4
S-90182 Umeå (Sweden)

Progress in Colloid & Polymer Science

Progr Colloid & Polymer Sci 70:101 – 108 (1985)

Rheological studies of sterically stabilised concentrated polystyrene latex dispersions under conditions of incipient flocculation

Th. F. Tadros

I.C.I. plc., Plant Protection Division, Jealott's Hill Research Station, Bracknell, Berkshire (U.K.)

Abstract: Measurements of shear stress-shear rate curves were made on concentrated aqueous polystyrene latex dispersions, stabilised with physically adsorbed poly(vinyl alcohol). Measurements were carried out under conditions of incipient flocculation, produced by addition of KCl or Na_2SO_4 at constant temperature (25 °C) or by raising the temperature of the dispersion, at constant electrolyte concentration. The results showed a change in the rheological behaviour of the dispersion from Newtonian to pseudoplastic (with hysteresis i.e. thixotropy) at a critical electrolyte concentration or at a critical temperature at constant electrolyte concentration. From plots of the extrapolated yield value versus electrolyte concentration and temperature (at constant electrolyte concentration) the critical electrolyte flocculation concentration (CFC) and the critical flocculation temperature (CFT) were established for the concentrated dispersion. The CFC of the concentrated dispersion was significantly lower than the corresponding value for a dilute dispersion. Above the CFC and CFT, the yield value τ_β showed an increase with increase of electrolyte concentration and with temperature. The results were quantitatively analysed using the elastic floc model proposed by Hunter and coworkers. Good agreement between the experimental yield values and those calculated from theory was obtained, thus confirming an applicability of the elastic floc model to the case of flocculating sterically stabilised dispersions.

Key words: Rheology, concentrated dispersions, sterically stabilised dispersions, polystyrene latex, incipient flocculation

Introduction

The flow behaviour (rheology) of suspensions, under conditions where strong attraction between the particles occurs, pose difficult problems both from theoretical and experimental points of view. This is due to the nonequilibrium nature of the resulting flocculated structure and the lack of quantitative information on the "microstructure" of the flocs formed. The "microstructure" depends on the nature of the flocculation process, on the one hand, and its evolution (which depends on the conditions) on the other. One of the earliest approaches towards a theoretical understanding of the non-Newtonian behaviour of flocculated systems was due to Goodeve [1] who recognised that the energy dissipation in the flow process could be subdivided into two main parts, namely the part due to the flow of fluid around the flow units (assuming these to be impenetrable) and the part due the attractive interactions between particles. This approach was further developed by Gillespie [2 – 5] who

could relate the Bingham yield value to the interaction energy between the flow units. Perhaps the most comprehensive investigations on the rheology of flocculated suspensions are those by Hunter and coworkers [6 – 14]. Such studies referred to systems flocculated in the deep energy minimum (sometimes referred to as coagulation) in the potential energy-distance curve described by the DLVO theory. Much of the behaviour of such coagulated suspensions could be explained by the so-called "elastic floc model" [6, 9, 10]. The basic flow unit in this model was considered to be an "elastic floc" which undergoes extension and compression during rotation in shear flow and during mutual collisions. The theory was intended to explain, in terms of the microrheological behaviour of the flow, the three characteristic flow parameters which describe the non-Newtonian behaviour of pseudoplastic flow of coagulated suspensions. These are the Bingham yield value, τ_β, the plastic viscosity, η_{PL}, and the critical shear rate $\dot{\gamma}_0$ above which the stress varies linearly with rate of shear [10]. The theories of Hunter et al. [6–14] showed that the most significant contributions to the energy dissipation in coagulated suspensions are the viscous energy dissipation due to the presence of the "spherical" flocs in the suspension and the energy dissipation during rupture of floc doublets. Clearly this model and the calculations presented are rather crude, but they describe, at least qualitatively, the flow behaviour of coagulated suspensions.

The above investigations are applicable to many practical systems, whereby destabilisation of a suspension is brought about by addition of electrolytes to the electrostatically stabilised suspension. However, many practical systems of the aqueous disperse type are stabilised by the addition of polymers that are either physically adsorbed to the particles or incorporated during their formation resulting in so-called steric stabilisation. Only a few investigations have been reported on the rheology of these relatively more complex systems. For example, Hunter et al. [15, 16] have investigated the flow behaviour of aqueous poly(methyl methacrylate) (PMMA) lattices stabilised by a block copolymer of poly(ethylene oxide-b-methacrylate) which was adsorbed onto the surface of the PMMA sphere by the smaller PMMA anchor groups of the copolymer. The authors used a simplified rheological approach, whereby the energy dissipation in a flocculating system was assumed to arise mainly from the rupture of doublets in a shearing field. Using this approach the authors were able to investigate the interaction properties of the flocculating dispersions. Interaction energy of the order of $5\,kT$ (where k is the Boltzmann constant and T the absolute temperature) were obtained for the PMMA system stabilised by adsorbed poly(ethylene oxide). More recently Tadros [17] studied the viscoelastic behaviour of concentrated aqueous polystyrene latex dispersions in the presence of physically adsorbed poly(vinyl alcohol) (PVA) as a function of surface coverage of the adsorbed polymer. At full coverage the viscoelastic behaviour was attributed to the combined action of long-range electrostatic repulsion and steric interaction due to the presence of long, "dangling" tails.

Of particular interest in sterically stabilised dispersions is the study of flow behaviour of such suspensions under the condition of incipient flocculation. This is obtained by reducing the solvency of the dispersion medium through the addition of sufficient electrolyte or by raising the temperature of the suspension, at constant electrolyte concentration [18]. The simplest case to consider is where the adsorbed polymer forms a "thick" layer (i.e. high molar mass) so that one can neglect any contribution from the van

der Waals attraction (G_A). Under conditions where the polymer solvent interaction parameter, χ, is greater than 0.5 (i.e. worse than θ-conditions), a significant value of $G_{\min}$ is obtained (see fig. 8 in ref. 18). It is of interest both theoretically and practically to study the rheology of such dispersions under these conditions and this constitutes the object of the present paper. The system studied was concentrated aqueous polystyrene latex dispersion with physically adsorbed PVA with a molar mass of 45,000. Incipient flocculation was induced by addition of KCl or Na_2SO_4 at constant temperature or raising the temperature at constant electrolyte concentration. The rheological behaviour of the system was investigated from shear stress-shear rate curves and in a few cases by measuring the modulus using a pulse shearometer.

Experimental

Materials

Water was doubly distilled from all glass apparatus. All other materials were analytical grade and used as received. Poly(vinyl alcohol) (Alcotex 88/10, supplied by Revertex Ltd., London) was the same sample used before [19]; it has a weight average molecular weight, M_w, of 45,000 and 12% acetate groups [19]. Two 50 litre batches of polystyrene latex were prepared using the method described by Goodwin et al. [20]; these will be referred to as lattices A and B. Both lattices were dialysed against distilled water until there was no further change in the conductivity of the dialysate. The lattices were concentrated by rotary evaporation, under vacuum. Latex A was concentrated to 37.2% w/w and it had an average particle radius of 163 ± 2 nm, whereas latex B was concentrated to 28.3% w/w and it had an average particle radius of 115 ± 2 nm. The particle size of the lattices was determined by using electron microscopy and a "Quantimet" image analyser.

Adsorption isotherms

The details of determining the adsorption isotherm of PVA on both latices has been described previously [19]. Both latices gave a high affinity isotherm with a saturation adsorption (plateau) value of 4.3 mg m^{-2}, which is in close agreement with the value obtained before for the same PVA sample on a similar latex [21].

Critical flocculation electrolyte concentration (CFC) of the dilute latex dispersions

Two methods were used to establish the critical flocculation electrolyte (KCl or Na_2SO_4) concentration of the dilute latex. In the first method, the turbidity, τ, of the dispersion (with mole fraction $\phi = 5 \times 10^{-4}$) was followed as a function of wavelength, λ, at various electrolyte concentrations at 25 °C, using an SP 1800 Pye-Unicam Spectrophotometer. Plots of $\log \tau$ versus $\log \lambda$ were linear over the wavelength range 400–600 nm. The gradient n of such a plot was then plotted versus electrolyte concentration [21]. At the CFC a strong reduction in n (due to increase in particle size from flocculation) was observed enabling the CFC to be accurately established. In the second method, the particle size was directly monitored using a Coulter Nanosizer (Coulter Electronics Ltd) as described previously [22]. The CFC of the dilute dispersions was found to be 1.27 mol dm^{-3} for KCl and 0.28 mol dm^{-3} for Na$_2$SO$_4$.

Rheological measurements

Shear stress-shear rate curves were carried out using a Haake-Rotovisco model RV100 (M.S.E. Scientific Instruments, Crawley, Sussex, U.K.) fitted with an M150 measuring head. The sensor system used was a co-axial cylinder type fitted with an MV1 bob having a radius of 20 mm. The radius of the stationary cup was 21 mm.

The shear rate was gradually and uniformly increased using the integral Haake programming unit. The shear stress-shear rate curves were plotted using the integral $x-y$ plotter.

The shear modulus, G_0, was determined using a pulse shearometer (Rank Bros., Bottisham, Cambridge, U.K.) based on the model originally described by van Olphen [23, 24] and later developed by Goodwin et al. [25, 26]. The instrument was placed in a constant temperature room kept at $20 \pm 0.5\,°C$. In this instrument the dispersion is placed between two parallel Perspex plates which are connected to piezoelectric crystals. A shear wave is generated at the bottom plate using a pulse generator which causes a small rotation (10^{-4} rad) of the upper plate and the shear wave velocity, u, through the dispersion is measured from a plot of the plate separation versus the time of propagation of the shear wave. The data are analysed on line using an Acorn Atom computer supplied with the instrument. G_0 is calculated from the expression,

$$G_0 = u^2 \varrho \tag{1}$$

where ϱ is the density of the dispersion.

Results

Influence of addition of electrolyte

Shear stress-shear rate curves of the latex dispersions (which were fully coated with PVA, i.e. at concentrations corresponding to the plateau of the isotherm) showed Newtonian behaviour below a critical electrolyte concentration, above which the flow curves became pseudoplastic, with some hysteresis (indicating thixotropy). From these pseudoplastic curves, the yield value, τ_β, was obtained by extrapolating the ascending curve to zero shear rate.

Figure 1 shows the variation of yield value, τ_β, with C_{KCl} at 25 °C, for a 30% w/w latex A dispersion. At $C_{KCl} < 0.95$ mol dm^{-3} the dispersions were virtually Newtonian showing no yield value or hysteresis in the shear stress-shear rate curves. For $C_{KCl} > 0.95$ mol dm^{-3}, the flow curves were typical of a

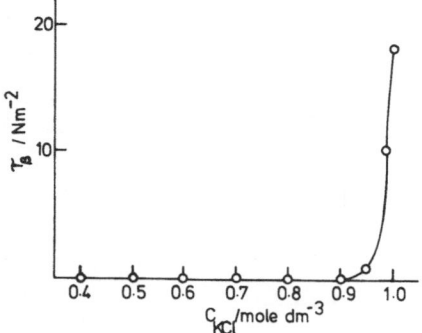

Fig. 1. Variation of extrapolated yield value with KCl concentration

thixotropic system, indicating flocculation. Thus, the CFC of the concentrated latex is about 0.95 mol dm^{-3}. This is lower than the CFC of the dilute latex (see experimental section) which is located at 1.27 mol dm^{-3}.

Figure 2 shows the variation of the extrapolated yield value with $C_{Na_2SO_4}$ for a 25% w/w latex B dispersion (note that this dispersion has a smaller particle radius than latex A and hence a smaller volume fraction was used for rheological measurements). The trend obtained is similar to that shown for KCl in figure 1. The shear modulus, G_0, measured using the pulse shearometer, is also shown as a function of $C_{Na_2SO_4}$ in figure 2. A measurable τ_β and G_0 is obtained above a critical value of $C_{Na_2SO_4}$, which in both cases is ~ 0.22 mol dm^{-3}. This electrolyte concentration is taken as the CFC for the concentrated latex,

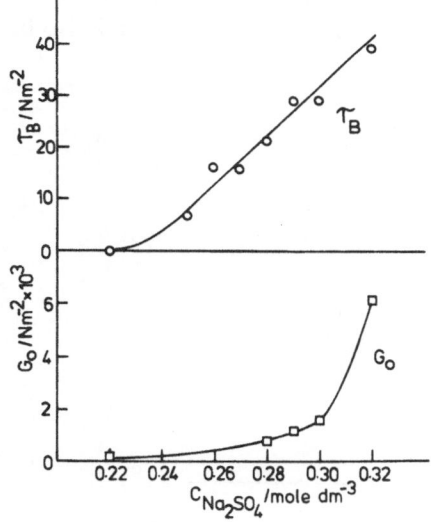

Fig. 2. Variation of extrapolated yield value and shear modulus with Na$_2$SO$_4$ concentration

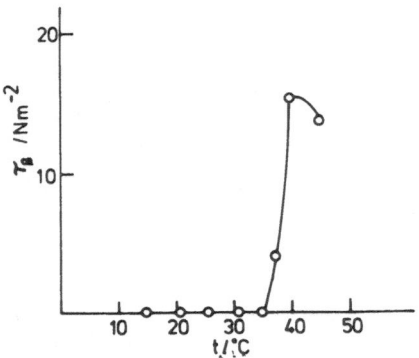

Fig. 3. Variation of τ_β with temperature at $C_{KCl} = 0.4$ mol dm^{-3}

which is also lower than that for the dilute dispersions (0.28 mol dm^{-3}). It should be noted that above the CFC, τ_β increases rapidly with increasing $C_{Na_2SO_4}$, whereas G_0 initially increases gradually with increasing $C_{Na_2SO_4}$, until $C_{Na_2SO_4} = 0.3$ mol dm^{-3}, above which there is a more rapid increase of G_0.

Influence of temperature

Figure 3 shows the variation of τ_β with temperature at $C_{KCl} = 0.4$ mol dm^{-3}. As can be seen from the figure τ_β is essentially zero until a critical temperature is reached, above which τ_β increases rapidly with increase in temperature, reaching a maximum above which there is a tendency of τ_β to fall again with further increase in temperature. The critical temperature corresponding to the abrupt increase in τ_β is 35 °C and this may be identified with the critical flocculation temperature (CFT) for this dispersion at this particular electrolyte concentration. It should also be

mentioned that below the CFT, the dispersion is essentially Newtonian, whereas above the CFT, the flow behaviour is that of a thixotropic system.

Figure 4 shows the variation of τ_β with temperature at two $C_{Na_2SO_4}$ values, namely 0.2 and 0.25 mol dm^{-3}. The trend observed is similar to that shown in figure 3 with KCl. The critical temperatures corresponding to the abrupt increase in τ_β are 20° and 25 °C for $C_{Na_2SO_4}$ equal to 0.25 and 0.20 mol dm^{-3} respectively. These temperatures may be identified with the CFT for these dispersions. Moreover, the maximum in τ_β above the CFT is still observed, particularly at 0.25 mol dm^{-3}.

Discussion

With the system investigated in this paper, namely polystyrene latex dispersion with adsorbed PVA chains, addition of KCl or Na$_2$SO$_4$, or increase in temperature at a given electrolyte concentration, leads to a reduction in the solvency of the chain. At a critical electrolyte concentration, or at a critical temperature (at a given electrolyte concentration), the mixing term in the steric interaction, ΔG_M, becomes zero since the χ-parameter (chain-solvent interaction parameter) becomes equal to 0.5. As mentioned in the introduction, with high molar mass polymers that give thick adsorbed layers, the CFC or CFT normally corresponds to the θ-point or θ-temperature of the chain in the medium under consideration. The present polymer, namely PVA, with a molar mass of 45,000 gives an adsorbed layer thickness of 47 nm [27]. Such a thick adsorbed layer is due to the presence of long, dangling tails [28 – 30]. It should be mentioned that, as a result of such a thick layer, the effective volume fraction, ϕ_{eff}, of the dispersion is much higher than the volume fraction of the particles ϕ_s. This can be readily seen from the following relationship between ϕ_{eff} and ϕ_s,

$$\phi_{eff} = \phi_s \left(1 + \frac{\delta}{a}\right)^3 \qquad (2)$$

where δ is the adsorbed layer thickness and a is the particle radius. Thus with latex A, $\phi_s = 0.3$ and $\phi_{eff} = 0.64$, whereas with latex B, $\phi_s = 0.25$ and $\phi_{eff} = 0.69$. Below the CFC or CFT, the dispersion is colloidally stable and shows Newtonian behaviour, as measured by steady-state shear stress-shear rate curves. However, creep measurements [17] showed a small residual modulus which was attributed, in part, to the interaction between the dangling polymer chains.

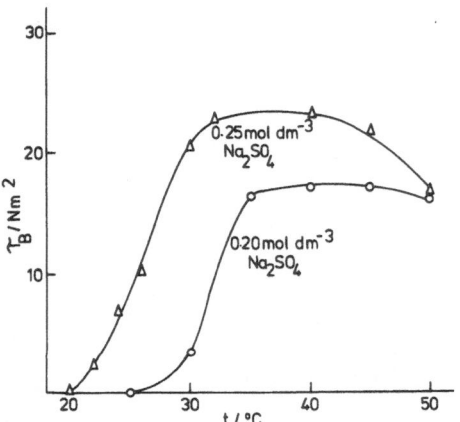

Fig. 4. Variation of τ_β with temperature at two Na$_2$SO$_4$ concentrations

As mentioned above at the CFC or CFT, $\Delta G_M = 0$ and any yield value measured should correspond to the residual van der Waals attraction. However, as we will see below, the contribution from the van der Waals attraction is small and, therefore, the yield value at the CFC or CFT is very small and unmeasurable. However, as we will also see, above the CFC or CFT, ΔG_M becomes negative and increases significantly with increasing electrolyte concentration and/or temperature leading to measurable τ_β values.

The energy arising from the van der Waals attraction may be calculated from the simple Hamaker equation, modified for the presence of an adsorbed layer, i.e.,

$$G_A = -\frac{(A_1^{1/2} - A_2^{1/2})^2 a}{12 H_0} \qquad (3)$$

where H_0 is the particle separation, A_1 is the Hamaker constant of the particle and A_2 that of the adsorbed polymer layer, which is given by

$$A_2 = [\phi_p A_p^{1/2} + \phi_m A_m^{1/2}]^2 \qquad (4)$$

where ϕ is the volume fraction of polymer in the adsorbed layer, ϕ_m that of the medium, A_p and A_m are the Hamaker constants of the polymer and medium (solvent) respectively. Using the following values, $A_1 = 19\,kT$, $A_p = 16.6\,kT$, $A_m = 9\,kT$ and $\phi_p = 0.067$ [27], G_A was calculated at a particle separation of $2\,\delta$ (i.e. 94 nm), which is the point at which the adsorbed layers touch. This gave a value of G_A of 0.24 and 0.17 kT for lattices A and B respectively. This corresponds to yield values of 0.013 and 0.024 Nm^{-2} respectively. These are certainly small and unmeasurable values. Thus, the point at which τ_β becomes measurable should correspond to the CFC or CFT value. Moreover, above the CFC or CFT, the contribution to τ_β from G_A is negligible and hence the main contribution to τ_β is from the G_M term, which now becomes attractive.

Comparison of CFC for concentrated and dilute dispersions

As mentioned above, the CFC of the concentrated suspension is significantly lower than that of the dilute dispersion. Indeed $(CFC)_{conc}/(CFC)_{dilute}$ is equal to 1.34 and 1.27 for KCl and Na$_2$SO$_4$ respectively. This difference may be attributed to one of the following reasons or a combination of both. The difference between $(CFC)_{conc}$ and $(CFC)_{dilute}$ may be accounted for if one considers the actual electrolyte con-

centration in bulk solution. With dilute dispersions ($\phi_s \sim 10^{-4}$) there is no difference between the nominal and actual electrolyte concentration since the particle volume fraction is very small. With the concentrated dispersion ($\phi_s = 0.3$ or 0.25), however, the electrolyte concentration was adjusted to the value required on the basis of the medium ($\phi_m = 0.7$ or 0.75) assuming the same electrolyte concentration in the adsorbed polymer layer (i.e. free draining polymer layer). Using the adsorbed layer thickness and amount of polymer adsorbed, the volume fraction of polymer in the adsorbed layer was calculated to be 0.067 for the present PVA sample [27]. If the adsorbed polymer layer is uniform, the electrolyte may be uniformly distributed in such a layer (corresponding to 6.7% PVA solution). However, recent studies using neutron scattering [31, 32] have shown that with PVA on polystyrene latex, the segment density shows two distinct regions; a very high density region close to the surface due to trains and loops, followed by a slow decay due to the presence of the tails which extends to the hydrodynamic thickness. If the assumption is made that the electrolyte is excluded from the more dense region of the adsorbed layer, i.e. the region of the trains and loops, then the actual concentration of Na$_2$SO$_4$ would be higher than the adjusted concentration due to this effect. Crude calculation shows that the difference between $(CFC)_{conc}$ and $(CFC)_{dil}$ could be accounted for if an excluded zone that is about half the adsorbed layer thickness is present around the polystyrene latex particles. However, it seems unlikely that an exclusion zone of this thickness is possible, since it is highly unlikely the thickness of the dense region reaches that magnitude. A more plausible alternative reason for the difference between $(CFC)_{conc}$ and $(CFC)_{dil}$ may be given if one considers the energetics of the flocculation process in more detail. This is given below.

Vincent et al. [33, 34] pointed out that the free energy of flocculation, ΔG_f, may be split into two contributions, i.e.,

$$\Delta G_f = \Delta G_i + \Delta G_{hS} \qquad (5)$$

where ΔG_i is the interaction free energy term which is a function of floc structure and depth of energy minimum, G_{min}, in the free energy-particle separation curves. ΔG_{hS} is the entropic contribution associated with the aggregation of hard spheres in the presence of other interparticle interactions. ΔG_{hS} is equal to $-T\Delta S_h$, where ΔS_h is the entropy of flocculation per particle. Clearly ΔS_h is negative and decreases in magnitude with increase in the volume fraction of particles, ϕ_S. Thus, since ΔS_{hS} is reduced in value in a concentrated dispersion, relative to its value in a dilute dispersion, it is clear that flocculation of a concentrated dispersion must occur at a lower G_{min} than for its dilute dispersion counterpart. Since G_{min} is proportional to electrolyte concentration, this results in a lower CFC for the concentrated dispersion compared to that for the dilute dispersion. However, the difference between $(CFC)_{conc}$ and $(CFC)_{dil}$ is likely to be due to the combined effects of unfavourable entropy term for concentrated dispersions and the exclusion effect due to the presence of dense layers.

Quantitative interpretation of the rheology results

For a quantitative description of rheological results, in particular the Bingham yield value, it is necessary to consider the total energy dissipated in the shear field. The total energy consists of three main contributions, namely from creep, network and viscous flow [15]. However, Neville and Hunter [15] considered that in a sterically stabilised dispersion, undergoing flocculation, the network structure is destroyed in the high shear region, such that the flow unit is the single colloidal particle and the largest floc unit is the doublet. Under these conditions, the extrapolated yield value is given by the expression,

$$\tau_\beta = \frac{3\,\phi_H^2}{\pi^2 (a+\delta)^3} E_S \qquad (6)$$

where ϕ_H is the hydrodynamic volume fraction of the particles, which is essentially ϕ_{eff} given by equation (2) and E_S is the energy required to separate a doublet which is the sum of the van der Waals and steric interactions (which is attractive under conditions of flocculation), i.e.

$$E_S = \left(\frac{Aa}{12H_0}\right) + G_S. \qquad (7)$$

As mentioned above, the contribution from the van der Waals attraction at separations corresponding to twice the adsorbed layer thickness is insignificant and, therefore, $E_S \sim G_S$. Thus, τ_β can be theoretically calculated from a knowledge of G_S and application of equation (6). Indeed, Neville and Hunter [15] attempted to interpret their τ_β versus temperature plots for sterically stabilised dispersions of PMMA particles in the presence of $MgSO_4$, using the above simple theory. However, in order to calculate G_S under flocculating conditions, it is necessary to know

the polymer-solvent interaction parameter, χ, or the expansion factor α (which is <1 above the CFC or CFT). Such information is not abailable for PVA in the electrolytes studied, namely KCl and Na_2SO_4. Experimental determination of the χ-parameter under worse than θ-conditions is not easy. Even if this is possible, calculation of G_S is not simple, without knowledge of the segment density distribution of the adsorbed layer. Moreover, the value of ϕ_H in equation (6) is a function of electrolyte concentration and temperature. This is due to the fact that δ decreases with increase of electrolyte concentration and/or temperature [15, 27]. Therefore, equation (6) can only be applied to give an approximate estimate of G_S from the measured yield value. Using this procedure, a value of G_S of the order of $300\,kT$ is obtained at $25\,°C$ in 1 mol dm^{-3} KCl and $200\,kT$ in 0.3 mol dm^{-3} Na_2SO_4. On the other hand, at constant electrolyte concentration, G_S increases with increase of temperature above the CFT. For example, in 0.4 mol dm^{-3} KCl, G_S increases from $\sim 70\,kT$ to $\sim 250\,kT$ as the temperature is increased from $37°$ to $50\,°C$. In 0.2 mol dm^{-3} Na_2SO_4, G_S increases from $\sim 25\,kT$ to $\sim 120\,kT$ as the temperature is increased from $30°$ to $45\,°C$. On the other hand, in 0.25 mol dm^{-3} Na_2SO_4, G_S increases from $\sim 15\,kT$ to $\sim 165\,kT$ as the temperature is increased from $22°$ to $40\,°C$.

Clearly the above floc rupture model is rather crude and, therefore, the values of G_S are only approximate, although they seem reasonable. Moreover, this simple floc rupture model may be used to explain the maximum in τ_β versus temperature (figs. 3 and 4). According to equation (6) $\tau = f(\phi_H^2) E_S$, where $f(\phi_H^2)$ is the collision frequency term. Although E_S increases with increase of temperature, $f(\phi_H^2)$ is a decreasing function of temperature as a result of decrease of solvency of the dispersing medium which leads to contraction of the adsorbed layer [15, 27]. The increase of E_S with increase of temperature initially outweighs any reduction of $f(\phi_H^2)$, but at higher temperatures, the reduction of $f(\phi_H^2)$ as a result of chain contraction may exceed the increase in E_S and this results in reduction in the measured τ_β.

A more realistic model for interpretation of the rheological results is the elastic floc model introduced by Hunter and coworkers [9, 10, 13]. In this model, the basic flow unit is considered to be an elastic floc which will undergo extension and compression during rotation in shear flow and during mutual collisions. The structural units (which persist at high shear rates) are assumed to be small flocs of particles (called floccules) which are characterised by the extent to which the particle structure is able to trap some of the dis-

Table 1. Parameters obtained using the elastic floc model

$C_{Na_2SO_4}/$ (mol dm^{-3})	$\eta_{PL}/(Nm^{-2})$ $\times 10^3$	$(\tau_\beta)_{exp}/$ (Nm^{-2})	C_{FP}	n_C $\times 10^{-13}$	$\dot{\gamma}_0/$ s^{-1}	F_M/N $\times 10^{-17}$	n_F $\times 10^{-8}$	F/N $\times 10^{-22}$	$a_{floc}/$ μm	$(\tau_\beta)_{theor}/$ (Nm^{-2})
0.25	10.8	6.8	1.63	1.79	152	1.7	4.40	4.2	24.3	6.6
0.28	32.7	21.4	1.84	1.16	211	5.2	1.71	7.6	39.2	25.9
0.29	33.8	29.0	1.90	1.15	352	5.3	1.99	8.2	36.2	33.8
0.30	33.0	29.0	1.89	1.16	421	5.2	3.41	15.2	27.7	36.5
0.37	31.6	–	1.88	1.17	–	5.0	4.19	1.79	25.0	–

persion medium. The floc density, or the amount of branching of particle chains within a floc, can be described by a quantity C_{FP} which is the ratio of the volume fraction of the flocs, ϕ_F, to the volume fraction of the particles, ϕ_S, i.e.,

$$C_{FP} = \phi_F/\phi_S . \tag{8}$$

Clearly the more open the structure of the floc, the larger is the C_{FP} value. ϕ_F may be calculated from the plastic viscosity, η_{PL}, using the Mooney equation (35),

$$\eta_{PL} = \eta_0 \exp\left(\frac{2.5\,\phi_F}{1 - k'\phi_F}\right) \tag{9}$$

where η_0 is the viscosity of the medium and k' is the so-called "crowding factor" which may be taken as 1.4 so that η_{PL} becomes infinity as ϕ_S approaches the close-packing value [14]. The values of C_{FP} using equations (8) and (9) are given in table 1 for flocculating dispersions using Na$_2$SO$_4$ (no calculations were made for KCl due to the limited number of points). It is clear that C_{FP} increases with increasing $C_{Na_2SO_4}$ above the CFC, indicating a more open structure with increase in electrolyte concentration.

To use the elastic floc model in a quantitative manner, it is necessary to have a model for the flocculated structure. For the present case of particles with adsorbed PVA chains, the most likely structure is that produced by interpenetration of the tails, under worse tha θ-conditions for the chains. A typical floc may be assumed to consist of strings of particles linked together in a more-or-less three dimensional network. The floc density (as measured by C_{FP}) is related to its strength by the number of chains, n_c, which pass through unit cross sectional area of floc [9, 10, 13]. n_c can be calculated from the total number of bonds per floc [10], i.e.

$$n_c = \frac{(C_{FP} + 0.7)}{b\,C_{FP}(C_{FP} - 1)\,a^2} \tag{10}$$

where b is a constant of the order of 10 which can be estimated from the close-packed structure. Values of n_c calculated using equation (10) are given in table 1.

The elastic floc model also suggests that,

$$C_{FP} = 1.5 + (F_M/b\,\eta_0 a^2) \tag{11}$$

where F_M is the maximum force of attraction between the particles. Values of F_M calculated using equation (11) are also given in table 1.

The critical shear rate, $\dot{\gamma}_0$, above which the $\tau - \dot{\gamma}$ curve becomes linear, is related to the number of floc-floc bonds, n_F, by the relation,

$$\dot{\gamma}_0 = \frac{n_F F_M}{5\,\eta_0} \tag{12}$$

values of n_F are also included in table 1.

The force required to break a floc doublet is then given by

$$F = (n_F/n_c)\,F_M . \tag{13}$$

Values of F are also included in table 1.

From n_F, the floc radius, a_{floc} can be calculated using the equation (13),

$$a_{floc} = (1.22\,\pi\,n_F)^{1/2} . \tag{14}$$

Values of a_{floc} are also included in table 1.

The yield value τ_β is given by (10)

$$\tau_\beta = \beta\lambda\eta_0(a_{floc})^2 \Delta\phi_S^2 C_{FP}/a^3 \tag{15}$$

where β is a constant ($= 27/5$) and λ is the orthokinetic capture efficiency which depends weakly on shear rate ($\lambda \propto \dot{\gamma}^{-0.18}$) and Δ is the distance through which bonds are stretched inside the floc by the shearing process. Thus, τ_β can be calculated using equation (15) provided reasonable values are used for $\dot{\gamma}$ and Δ. $\dot{\gamma}$ may be taken to be equal to $\dot{\gamma}_0$, and a reasonable value of Δ would be 0.5 nm [10]. In this manner, a comparison between τ_β calculated from equation (15) and the experimental τ_β may be made. The results are also given in table 1. Given all the approximations made, it is clear that the agreement between experimental and theoretical values of τ_β is quite good, confirming the validity of the elastic floc model for flocculating sterically stabilised dispersions.

Conclusions

Sterically stabilised concentrated dispersions under conditions of incipient flocculation give flow behaviour (rheology) which may be described by the elastic floc model suggested by Hunter and coworkers [9, 10, 13] for coagulated suspensions. In this model the basic units of flow are elastic flocs consisting of a certain number of flocculi. Such flocs are compressed and extended during rotation in shear flow and during mutual collisions. With increase of extent of flocculation, e.g. by increasing electrolyte concentration above the CFC or increasing the temperature above the CFT, the floc structure becomes more open entrapping more of the dispersion medium. The elastic floc model gives a measure of the yield values, with only one adjustable parameter, namely the distance, Δ, through which bonds are stretched inside the floc by the shearing process. With the present system, the theoretical τ_β values agree well with the experimental values when a value of $\Delta = 0.5$ nm is assumed.

Acknowledgements

Most of the experimental work described in this paper was carried out by Mr. J. M. Cook (Salford University) and Miss Sheila Spence (Brunel University) during a six-month industrial training period at Jealott's Hill. I am indebted to Mr. D. Heath for preparing the latex and supervising part of the experimental work. Permission of the Management to publish this work is also appreciated.

References

1. Goodeve CF (1939) Trans Faraday Soc 35:342
2. Gillespie T (1960) J Colloid Sci 15:219
3. Gillespie T (1966) J Colloid Interface Sci 22:563
4. Gillespie T (1976) IN: Kerker M (ed) Colloid and Interface Science, Vol IV. Academic Press, New York, p 465
5. Gillespie T (1983) J Colloid Interface Sci 93:337; 94:166
6. Friend JP, Hunter RJ (1971) J Colloid Interface Sci 37:548
7. Frith BA, Hunter RJ (1976) J Colloid Interface Sci 57:248
8. Frith BA (1976) J Colloid Interface Sci 57:257
9. Frith BA, Hunter RJ (1976) J Colloid Interface Sci 57:266
10. van de Ven TGM, Hunter RJ (1977) Rheol Acta 16:534
11. Hunter RJ, Frayne J (1979) J Colloid Interface Sci 71:30
12. Hunter RJ, Frayne J (1980) J Colloid Interface Sci 76:107
13. Hunter RJ (1982) Advances Colloid Interface Sci 17:197
14. Hunter RJ, Matarase R, Napper DH (1983) Colloid and Surfaces 7:1
15. Neville P, Hunter RJ (1974) J Colloid Interface Sci 49:204
16. Frith BA, Neville PC, Hunter RJ (1974) J Colloid Interface Sci 49:214
17. Tadros ThF (1984) ACS Symposium Series No 240:411
18. Tadros ThF (1982) Polymer Adsorption and Dispersion Stability. In: Tadros ThF (ed) The Effect of Polymers on Dispersion Properties. Academic Press, London
19. Garvey MJ, Tadros ThF, Vincent B (1974) J Colloid Interface Sci 49:57
20. Goodwin JW, Hearn J, Ho CC, Ottewill RH (1974) Colloid and Polymer Sci 252:464
21. Tadros ThF, Vincent B (1972) J Colloid Interface Sci 72:505
22. Tadros ThF, Vincent B (1980) J Phys Chem 84:1575
23. Van Olphen H (1956) Clay Clay Minerals 4:68
24. Van Olphen H (1958) Clay Clay Minerals 6:106
25. Goodwin JW, Smith RW (1974) Faraday Disc Chem Soc 57:126
26. Goodwin JW, Khider AH (1976) In: Kerker M (ed) Colloids and Interface Sci, Vol IV, p 529
27. Van den Boomgaard Th, King TA, Tadros ThF, Tang H, Vincent B (1968) J Colloid Interface Sci 61:68
28. Von Vliet T, Lyklema J (1978) Disc Faraday Soc 65:25
29. Cain FW, Ottewill RH, Smitham JB (1978) Disc Faraday Soc 65:33
30. Sonntag H, Emke B, Miller R, Knapschinsky L (1982) In: Tadros ThF (ed) The Effect of Polymers on Dispersion Properties. Academic Press, London, p 207
31. Barnett KG, Cosgrove T, Crowley TL, Tadros ThF, Vincent B (1982) In: Tadros ThF (ed) The Effect of Polymers on Dispersion Properties. Academic Press, London, p 183
32. Cosgrove T, Crowley TL, Vincent B, Barnett KG, Tadros ThF (1981) Faraday Symposium (Chemical Society) 16:101
33. Cowell C, Li-In-On R, Vincent B (1978) J Chem Soc, Faraday I 74:337
34. Vincent B, Luckham PF, Waite FA (1980) J Colloid Interface Sci 73:508
35. Mooney M (1951) J Colloid Sci 6:162

Received July 7, 1984;
accepted October 15, 1984

Author's address:

Dr. Th. F. Tadros
I.C.I. plc.,
Plant Protection Division
Jealott's Hill Research Station
Bracknell
Berkshire
RG12 6EY (U.K.)

Progress in Colloid & Polymer Science   Progr Colloid & Polymer Sci 70:109 – 112 (1985)

Characterization of the surface of cellulosic fibers using aqueous two-phase partitioning

L. Ödberg[1], G. McBride[1], and K.-E. Magnusson[2]

[1]Institute for Surface Chemistry, Stockholm (Sweden)
[2]Dept. of Medical Microbiology, University of Linköping, Linköping (Sweden)

Abstract: The physical chemical surface properties of cellulosic fibers have been assessed using aqueous two-phase partitioning in polyethyleneglycol (PEG) – dextran systems containing hydrophobic palmitoyl-PEG and positively charged trimethylamino-PEG. The sizing with alkylketenedimer (AKD) of bleached kraft pulp could easily be monitored with this new technique. The adsorption of oleate on thermomechanical pulp could also be followed. In this case the oleate adsorbs preferentially on the hydrophobic parts of the fibers making them more hydrophilic. We also investigated bleached sulphite fibers treated in various ways. Ethanol extraction made the fibers more hydrophilic, obviously by removing resinous material from the fiber surfaces. Heat treatment of the fibers made the fibers more hydrophobic. The resinous material probably melts and covers a larger part of the fiber surface.

Key words: Cellulose, adsorption, two-phase partitioning

Introduction

It is of great interest to characterize the surface properties of cellulosic fibers, e.g. in connection with sizing operations, recycling of fibers and the use of cellulosic fibers in composite material. Information on the chemical composition of the fiber surface can often be obtained by ESCA-spectroscopy, both for unmodified [1] and chemically modified [2, 3] fibers. The surface energy can, in principle, be obtained by measuring the contact angle. Attempts have been made to measure the contact angle of individual fibers using a micro balance technique [4, 5]. Other methods to characterize the surface energy are based on the rate of adsorption in sheets of paper [6]. These tests are however also sensitive to the structure of the sheet. The contact angles of small particles can also be obtained by studying the engulfment or rejection of particles at solidification fronts. This technique has been used for small polymer particles [7, 8].

Aqueous two-phase partitioning in polyethylene glycol (PEG) – dextran involves the partition of macromolecules or particles between one phase rich in PEG and one phase rich in dextran. Partition in this system is determined by the surface properties of the particles. In general hydrophobic material is partitioned towards the PEG phase and the hydrophilic

material towards the dextran phase. The method is described in detail in a monograph by Albertsson [9]. In order to isolate the effect of hydrophobicity some of the PEG molecules are replaced by PEG esterified with palmitoylchloride (palmitoyl-PEG = P-PEG) [10]. The hydrophobicity can now be assessed by comparing the distribution in the basal system and the system with palmitoyl-PEG. A hydrophobicity index can be constructed as described below in Materials and methods.

Aqueous two-phase partitioning has been extensively used for the separation and the characterization of proteins [9], cellular organelles [11], bacteria [12] and animal cells [13, 14]. In addition to separation according to hydrophobicity the surface charge of the particles can be used. In this case some of the P-PEG molecules are replaced by positively charged trimethylamino PEG or negatively charged sulfonate-PEG [15]. For cellulosic fibers the charge can in most cases be satisfactorily determined using electrophoretic techniques and we have therefore concentrated on the hydrophobic/hydrophilic aspect of partitioning, although there might very well be interesting information to gain from partitioning according to charge.

To our knowledge the phase partition method has not been used to characterize the surface of cellulosic fibers. In the present article it will be shown that this is indeed possible and that the method offers many interesting possibilities.

Materials and methods

Two-phase partitioning

The procedure to obtain the aqueous two-phase systems used has been described elsewhere [16]. The basal system contained 4.4% (w/w) polyethylene-glycol 6000 (PEG; Carbowax 6000, Union Carbide, New York, N.Y.) and 6.2% (w/w) dextran T500 (Pharmacia Fine Chemicals AB, Uppsala, Sweden) in 0.03 M tris-hydroxymethylamino-methane buffer (tris), pH 7.0. After equilibrating the two-phase system overnight at 4°C in a separation funnel, the bottom phase (rich in dextran) and the top phase (rich in PEG) were collected and stored separately at 4°C. For the partitioning studies, 2 ml of the bottom and 2 ml of the top phase were pipetted into graduated (volume) test tubes. To assess hydrophobic interaction and negative charge, 0.2 ml 5 or 10% (w/w) palmitoyl-PEG (0.13 mmol palmitic acid per gram) or 0.2 ml 5% (w/w) bis-trimethyl-amino-PEG (TMA-PEG) in phosphate buffered saline (PBS; pH 7.3) was added. In the basal system the same amount of PEG was added. 0.2 ml of the suspension of cellulosic fibers was then added to the test tubes with the different two-phase systems, and the tubes inverted (twenty times) for mixing. They were then kept at 4°C for 30 min for separation of the phases. After determining their volumes 1.0 ml aliquots were withdrawn from each of the two phases. The rest was mixed with a Vortex homogenizer and 1.0 ml was withdrawn. This sample thus also contained material adhering to the interface. The samples were used to determine the distribution of fibers. The variation between two experiments was less than 5%. To obtain consistent results it is necessary to take certain precautions. The pulp samples should be taken at one time and used directly since the surface properties change both for dry (selfsizing) and wet pulp. Sodium azid was added to the pulp dispersions to avoid bacterial growth. Furthermore, it is essential to work with freshly prepared polymer solutions and to keep the temperature strictly controlled.

The results are expressed by the percentage material in the top (T) and bottom (B) phase, the remainder making 100%, is found at the interface (I). The change of the distribution by the addition of TMA-PEG or P-PEG relative to the distribution in the basal system, is calculated as accumulation of new material in (T) plus removal of material from (B). This index consists of three components: the transfer of material (i) from (B) to (I), (ii) from (B) to (T) and (iii) from (I) to (T). If particles are unaffected by ligand PEG, the index equals zero; it becomes 200 if the entire particle population is moved from a position in (B) in the basal system to (T) in the system containing ligand PEG. This index was chosen so as to monitor all three of the components above. Any index based solely on the transfer to (T) ignores component (i), while one based on the transfer of particles from (B) would ignore component (iii).

Counting of cellulosic fibers

The cellulosic fibers in the samples were counted by using instruments based on the electrozone sensing technique. For the experiments with the fines fraction (see below) we used a Coulter Counter Model ZF (Coulter Electronics, South Dunstable, Beds., England) connected to a Channelyzer C-1000 (Highleah, Fl., USA). The medium size fibers were counted by using an Elzone Particle Counter (Particle Data, Elmhurst, Ill., USA). In both cases a test tube with 100 µm-aperture was used. These instruments also give the size distribution of the fibers counted.

Cellulosic fibers

Intact cellulosic fibers have a length of 1 – 3 mm and a width of 30 – 40 µm. In a commercial pulp there are fiber fragments of all possible lengths. As will be discussed below the very long fibers are not suitable for partitioning experiments. We therefore fractionated the fibers using the standard technique in a Dynamic Drainage Jar [17]. In the first fractionation we used a 125 P screen (0.08 mm holes). The fraction that passed this screen is denoted *fines* fraction. The fraction that was witheld is called the *long fiber* fraction. We further prepared from the long fiber fraction a *medium* fiber fraction by taking the fibers of the long fiber fraction that passed a 100 P screen (0.1 mm holes). Of course with fibrous material the hole size only gives a very rough idea of the size of the fibers in the various fractions. For the medium fraction the majority of the fibers were 0.1, 0.2 mm long even though there were fibers up to 0.5 mm.

Results

Fiber sizes

It can be shown, all other factors being equal, that the larger a particle, the larger the tendency to accumulate at the interface [9]. It was therefore not too surprising to find that when using the long fiber fraction all the fibers were found at the interface forming what looked like a separate phase. The height of this "phase" i.e. its water content did vary with the type of pulp used. We have so far not tried to use this phenomenon quantitatively. However both the fines fraction and the medium fraction could be partitioned in the system. For the medium fraction a rather high percentage was often found at the interface. When working with the medium fraction we also observed that fibers originating from samples containing the interface had a slightly larger average size than fibers in the samples from the top or bottom phases.

P-PEG concentration

We have tried to find a suitable level of P-PEG by adding 5 and 10% P-PEG. The results given below in table 1 are for medium size fibers of unbleached softwood kraft pulp.

Both levels of P-PEG work although 10% gives a higher hydrophobicity index. Our experience is that

Table 1. Partition of medium size fibers, unbleached softwood kraft pulp. Results in percent

	(T)	(I)	(B)
Basal system	25	70	5
5% P-PEG	34	53	13
10% P-PEG	38	53	9

the higher level is preferable although both levels have been used in the present study.

Repeated partitioning

One question that naturally arises is: Do the fibers in our samples have the same surface properties and partition between the phases just like a well defined chemical species partition between two phases? Or do the fibers that are found in the top phase have different properties from those in the bottom phase? To investigate this question we have distributed medium size bleached softwood fibers in the usual way in the basal system. We have then collected fibers from the bottom phase and partitioned them again. The results are given in table 2.

As can be seen the distribution is almost the same the second time. This indicates that the fibers in this very sample are rather homogeneous and distribute statistically between the phases weighted according to their affinity for the phases. Had the sample consisted of various fractions a repeated distribution would not give the same results. For heterogeneous samples repeated distribution of course gives the possibility for fractionation. This is the basis for much of the use of this method in the field of cell biology.

AKD-sized fibers

Cellulosic fibers can be sized i.e. given hydrophobic properties in various ways. The classical method is using rosin size with the addition of alum. This method does not work for neutral pH where one instead reacts the fibers with alkylketene dimer (AKD).

We have sized a bleached softwood kraft pulp with AKD. We have taken a sample from this sized pulp.

Table 2. Repeated partitioning of medium size bleached softwood fibers. Basal system. Results in percent

	(T)	(I)	(B)
First partitioning	4	26	70
Partitioning of bottom phase	3	30	67

Table 3. Partition of medium size bleached softwood fibers. AKD-sized to various degrees. Distribution in percent

	(T)	(I)	(B)
B0	4	26	70
P0	3	30	67
B1	18	25	57
P1	26	39	35
B2	13	59	28
P2	21	65	14
B3	12	79	9
P3	35	55	10

The rest of the pulp has been reslushed and sized again. In total three sizings were performed. The reason for repeating the treatment is that it is believed that in one sizing operation only a rather small fraction of the fiber surface is reacted [18]. From the pulp sample taken we prepared medium size fiber samples. The results of the partition studies are given in table 3. B stands for basal system, P for system with 0.2 ml 10% P-PEG. 0 stands for unsized fibers, 1 for sized once etc.

As can be seen the unsized fibers are hydrophilic and heavily partitioned towards the dextran phase. Addition of P-PEG does not influence this partitioning very much. However, when the fibers are sized they are removed from the dextran phase both in the basal system and even more so in the systems with added P-PEG. After three sizings only 10% of the fibers remain in the dextran phase.

Adsorption of sodium oleate on TMP-fibers

In these experiments various amounts of sodium oleate were added to a pulp suspension of TMP-fibers in a Dynamic Drainage Jar at pH 10. After 10 minutes we added to some of the samples $CaCl_2$ to $10\,dH^0$ ($= 1.8$ mM). After another 30 minutes we collected the fines fraction from this pulp suspension and partitioned the fines in our systems. In what follows we report the hydrophobicity index ($=$ accumulation in top phase + depletion in bottom phase relative to the basal system on addition of P-PEG) since this index tends to vary less than the actual distributions. We also made experiments with adding positively charged TMA-PEG to the PEG phase. We report the corresponding charge index which is positive for negatively charged fibers. The results are given in table 4.

As can be seen an increased amount of sodium oleate makes the fibers less hydrophobic. Obviously

Table 4. Hydrophobicity index and charge index for TMP-fines with adsorbed sodium oleate. 0.2 ml 5% P-PEG and 0.2 ml 5% TMA-PEG

	Hydrophobicity index	Charge index
50 mg/l NaOl	56	58
50 mg/l NaOl + 10 dH^0 Ca^{2+}	29	54
100 mg/l NaOl	51	66
100 mg/l NaOl + 10 dH^0 Ca^{2+}	44	57
200 mg/l NaOl	47	64
200 mg/l NaOl + 10 dH^0 Ca^{2+}	45	58
300 mg/l NaOl	37	19
300 mg/l NaOl + 10 dH^0 Ca^{2+}	22	28

the oleate hydrophilizes hydrophobic parts of the TMP fibers. The addition of Ca^{2+} ions has only a minor effect on the hydrophobicity. The Ca^{2+} ions cannot compete for the already adsorbed oleate ions except perhaps for the lowest addition of oleate. The charge of the fibers does not change very much according to our experiments with TMA-PEG. Too much significance should not be attached to the experiments at 300 mg/l NaOl since these results show much greater uncertainties ($\pm 15\%$) than any of the other experiments.

Adsorbed pitch

All fibers, even bleached ones, seem to have some resinous, hydrophobic material on their surfaces. This material is usually referred to as pitch. We have investigated a bleached sulphite pulp. To remove some of the pitch the pulp can be extracted with ethanol. If, on the other hand, the pulp is heat treated (105 °C; 3 h) the pitch is believed to spread over the surface. We made partition experiments for fines fractions from these pulps. A bleached sulphite pulp is of course hydrophilic and all samples partition towards the dextran phase in the basal system. On addition of 0.2 ml 5% P-PEG the hydrophobicity index is 0 for the untreated fiber, for the ethanol extracted fines it is -9 and for the heat treated $+5$. These results are thus entirely in accordance with our expectations.

Discussion

The aim of this work has been to show that the phase partition technique has great potentials for characterizing cellulosic fibers. The results above clearly show that the method gives a good measure of

the hydrophobicity of the surface. It is quite probable that the procedure can be improved by various modifications such as changing concentrations or by using other fiber fractions. As described by Albertsson [9] in his monograph there also exists the possibility to use other basal systems than PEG/dextran.

So far we have not been able to use the long fiber fraction for any quantitative measurements. As indicated above, this might not be quite out of reach. However, we feel that there is enough intact fiber surface on the medium size fibers for a correct evaluation of surface properties.

Acknowledgements

We should like to thank Per Stenius and Tom Lindström for stimulating discussions and Gunnel Söderberg for making the AKD sizings.

References

1. Dorris GM, Gray DG (1978) Cell Chem Techn 12:735
2. Takeyama S, Gray DG (1982) Cell Chem Techn 16:733
3. Ödberg L, Ström G (1983) Svensk Papperstidn 86:R141
4. Young RA (1976) Wood and Fiber 8:120
5. Klungness JH (1981) TAPPI 64:65
6. One such test is: SCAN-Test P 12:64
7. Neumann AW, Omenyi SN, van Oss CJ (1979) Colloid and Polymer Sci 257:413
8. Omenyi SN, Smith RP, Neumann AW (1980) J Colloid Interface Sci 75:117
9. Albertsson P-Å (1971) Partition of Cell Particles and Macromolecules 2nd ed. Almqvist & Wiksell, Uppsala, and John Wiley & Sons, New York
10. Johansson G (1970) Biochem Biophys Acta 222:381
11. Eriksson E, Johansson G (1979) In: Reid E (ed) Cell Populations, Methodological Surveys (B). Biochemistry: Vol 9. Ellis Horwood Ltd, Chichester, pp 81 – 90
12. Magnusson K-E, Johansson G (1977) FEMS Microbiol Lett 2:255
13. Walter H (1977) In: Catsimpoolas N (ed) Methods of Cell Separation, Vol 1. Plenum, New York, pp 307 – 354
14. Dahlgren C, Kihlström E, Magnusson K-E, Stendahl O, Tagesson C (1977) Exptl Cell Res 108:175
15. Johansson G (1970) Biochem Biophys Acta 222:381
16. Kihlström E, Magnusson K-E (1980) Cell Biophysics 2:177
17. Britt KW (1973) TAPPI 56:83
18. Lindström T, personal communication

Received July 6, 1984;
accepted September 5, 1984

Authors' address:

Dr. Lars Ödberg
Swedish Forest Products Research Laboratory
Box 5604
S-11486 Stockholm (Sweden)

Progress in Colloid & Polymer Science   Progr Colloid & Polymer Sci 70:113 – 118 (1985)

Flow patterns of immiscible liquid/fluid displacement in a capillary tube

É. Kiss, J. Pintér, and E. Wolfram

Department of Colloid Science, Loránd Eötvös University, Budapest (Hungary)

Abstract: Liquid flow as disturbed by the presence of a fluid interface of definite shape was studied during immiscible liquid/fluid displacement in a capillary tube. Microscopic cinematography with tracer method has been used to obtain flow patterns on both sides of the moving interface. The effect of the curvature of the interface on the distribution of velocity components was investigated. The experimental results obtained for flat liquid/fluid interfaces were compared with those calculated from an approximate solution of the flow equation by Kafka and Dussan V.

Key words: Liquid/fluid displacement, capillary flow, flow pattern, shape of the interface

Introduction

Fluid/liquid/solid adhesion has a predominant role in a great number of natural and industrial processes, and it is widely investigated. Displacement of the three-phase contact line usually takes place in both the formation and the ceasing, and the modification of the adhesion. The essential problem is how the contact line can move and what the mechnism of the displacement is.

Wetting under static conditions can experimentally be well characterized only in a few geometrically simple systems. However, dynamic wetting behaviour is not clearly defined and obtainable with difficulty. The connection between static contact angle, velocity of displacement, capillary number, geometry of the system as independent variables, and dynamic contact angle as a dependent variable has been investigated in a number of fluid/liquid/solid systems. In certain cases where the contact line does not move along with the bulk liquid, the shape of the interface is more important than the dynamic contact angle. The distortion of the fluid interface can result in such phenomena as finger-forming, film-imbibition, liquid-entrapment, etc.

Two-phase flow in a capillary tube in its entirety and parameters influencing the efficiency of oil displacement were studied in our previous works [1, 2]. The mechanism of the contact line motion is focused upon in our present experiments dealing with immiscible liquid/fluid displacement.

Considering the difficulty of observing the shape of the moving interface in the very vicinity of the contact line, the flow conditions on both sides of the interface, which demonstrate the effect of the contact line motion in visible dimensions, were investigated. The flow of two liquid phases contacting each other along one fluid interface in a capillary tube was carried out at a constant velocity. Velocity fields of both liquids were determined simultaneously at the moving interface in order to obtain some information about the effect of the motion of a meniscus on tube flow, its extension and the flow behaviour of the fluid interface itself.

Preliminary considerations

A contradiction between moving of the contact line and of the bulk liquid far from the contact line can easily be shown in the case of a two-liquid capillary flow (fig. 1). This model system has proved to be suitable for theoretical calculations too, because of its symmetry. It is well known that a liquid flow in a tube at a low Reynolds number is a laminar flow with parabolic velocity distribution and with zero velocity at the capillary wall (fig. 1a). On the other hand, it is also an empirical fact that displacement does occur,

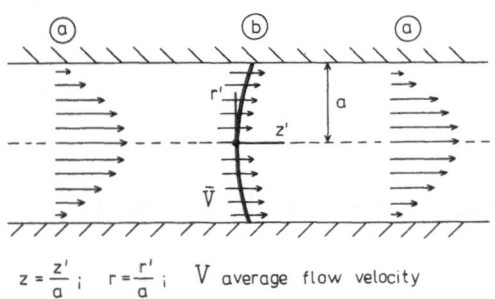

$$z = \frac{z'}{a} \; ; \quad r = \frac{r'}{a} \; ; \quad \bar{V} \; \text{average flow velocity}$$

Fig. 1. Two-liquid flow in a capillary tube. Velocity distribution at the interface (b) and away from it (a)

and the contact line does move along the solid surface. Every part of the interface moves with the same velocity as its shape is unchanged during the flow under steady-state conditions (fig. 1b). This kind of flow has to transform into the Hagen-Poiseuille flow in the transition regions at the interface. Radial flow is required by the transformation in these regions.

The supposed flow patterns of a "pure" tubular flow and of that with a fluid/liquid interface as demonstrated by streamlines are shown in figure 2a and 2b, respectively. (The coordinate system is fixed to the interface moving together with it. The meniscus seems to stay still and the capillary wall to move at the mean velocity in the opposite direction.) The streamlines are straight lines parallel to the capillary wall in the case of the Hagen-Poiseuille flow (fig. 2a), whereas curved streamlines are formed in the presence of an interface (fig. 2b). The latter are directed towards the wall and the tube axis in the advancing and in the receding liquids, respectively.

Certain aspects of this type of motion have already been detected experimentally, e.g., radial motion towards the contact line in the advancing liquid behind a convex meniscus was indicated by a dye tracing method [3]. The mere presence or absence of radial flow near the interface in the advancing liquid was brought into connection with the formation of a con-

tinuous liquid film on the capillary wall against the contact situation of liquids [4]. The sole work of Karnis and Mason [5] gave a detailed analysis of the flow conditions behind a flat air/liquid meniscus.

As the tube flow is at variance with the motion of the interface, analytic solution of the Navier-Stokes equation cannot be obtained considering the no-slip condition [6, 7]. To overcome this difficulty, the existence of a narrow region near the contact line is supposed in most of the cases where the no-slip condition is replaced by the classical slip or free slip conditions [8]. The approximative calculations, making use of different boundary conditions in the two regions of the system, can provide a complete solution.

For a two-liquid system Kafka and Dussan V. [9] gave an approximate solution of the Navier-Stokes equation which was assumed to be satisfied within both fluids:

$$CaRe^K \left(v_r^K \frac{\partial v_r^K}{\partial r} + v_z^K \frac{\partial v_r^K}{\partial z} \right)$$

$$= -\frac{\partial p^K}{\partial r} + \bar{\mu}^K Ca \left[\frac{\partial}{\partial r} \left(\frac{1}{r} \frac{\partial}{\partial r} (r v_r^K) \right) + \frac{\partial^2 v_r^K}{\partial z^2} \right]$$

$$CaRe^K \left(v_r^K \frac{\partial v_z^K}{\partial r} + v_z^K \frac{\partial v_z^K}{\partial r} \right)$$

$$= -\frac{\partial p^K}{\partial z} + \bar{\mu}^K Ca \left[\frac{1}{r} \frac{\partial}{\partial r} \left(r \frac{\partial v_z^K}{\partial r} \right) + \frac{\partial^2 v_z^K}{\partial z^2} \right]$$

where K denotes the advancing and receding fluids (A and R), v_r and v_z are the radial and axial velocity components, p is the pressure.

The continuity equation

$$\frac{1}{r} \frac{\partial}{\partial r} (r v_r^K) + \frac{\partial v_z^K}{\partial z} = 0$$

and the following boundary conditions were used for the solution:
– Bond-number $= \dfrac{\varrho g l^2}{\gamma}$, Capillary-number $= \dfrac{v \mu}{\gamma}$,

Reynolds-number $= \dfrac{v a \varrho}{\mu}$ are small, and the shape

of the interface is near flat. (l is the characteristic length of the system, ϱ, μ, v are density, viscosity and velocity of the liquid, a is the tube radius, γ is the fluid interfacial tension.)
– no slip condition is valid along the moving wall except for the region of the fluid interface
– $v_z = 0$ at $z = 0$
– $v_{r,A} = v_{r,R}$ at $z = 0$

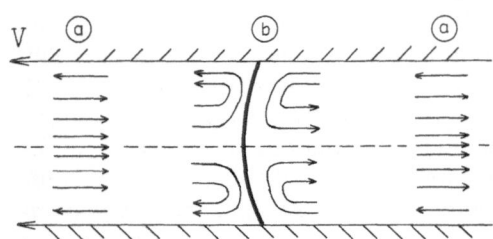

Fig. 2. Flow patterns of tube flow (a) and of two-liquid flow (b)

– tangential stress exerted by either fluid upon the meniscus is the same at $z = 0$

– capillary pressure of the interface is balanced by the normal stress difference.

Stream function values, ψ^K, were obtained which can be used to estimate velocities at any point within the two fluids:

$$v_r^K = \frac{1}{r} \frac{\partial \psi^K}{\partial z}; \qquad v_z^K = -\frac{1}{r} \frac{\partial \psi^K}{\partial r}.$$

Hence theoretical results [9] can be compared with our experimental results only for the case of a flat interface.

Experimental

Materials

Cylindrical capillaries with a diameter of 0.45 mm were made of pyrex glass by a Hewlett-Packard capillary drawer apparatus. Its surface was hydrophobized partly or completely by sylilation to form fluid interfaces with different shapes of the same fluids and at the same velocity (2 mm/s).

Air, water and *n*-dodecane (Fluka, pract., viscosity$_{298\,K}$: 1.35 mPas) were used to compose the fluid/liquid and liquid/liquid systems.

In order to visualize the flow patterns small particles with the required size, shape, density and wetting properties dispersed in the liquid phases were chosen as tracer materials. Polystyrene lattices (diameter: $10 - 15$ μm) and polyethylene particles (diameter: $15 - 25$ μm) proved to be suitable for this purpose in the water and in the *n*-dodecane, respectively.

Equipment

The capillary tube placed into a rectangular glass cell was surrounded by an appropriate oil mixture with the same refractive index as that of the tube to eliminate the undesired reflection and optical distortion. The uniformity of the curvature of the glass capillaries was checked microscopically looking at several short pieces of the tube from their axial directions.

The application of a microscope objective enables us to observe the liquid flow with the moving tracer particles only in the plane of the axial section of the capillary. The two-phase flow produced by the same device as used in our previous work [2] was studied by high-speed cinematography. The flow phenomena were analysed by determinating the position of the tracer particles on the frames of the film one using a film analysator apparatus of NAC type. This method produces the pathlines of particles which are equivalent to streamlines in the case of a steady-state flow.

Results

Two frames taken from the film show the tracer particles near the moving liquid/liquid interface (figs.

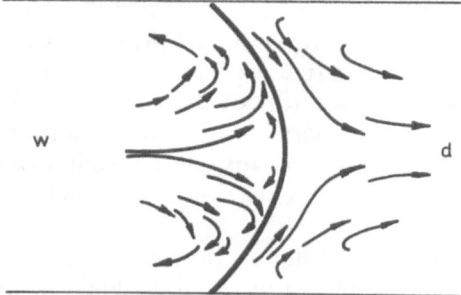

Fig. 3. Two-phase flow with tracer particles within both liquids; flow patterns of displacement of *n*-dodecane by water

4a

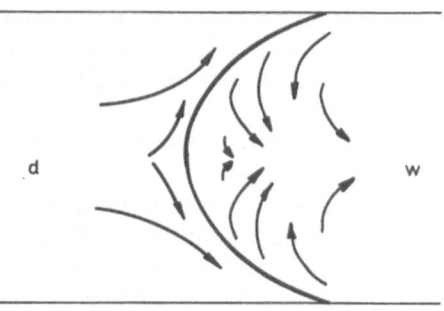

4b

Fig. 4. Two-phase flow with tracer particles within both liquids; flow patterns of displacement of water by *n*-dodecane

3a, 4a). The advancing liquid is water and the receding one is *n*-dodecane in figure 3, and the sequence of the liquids is reversed in figure 4. Flow patterns determined from the displacement of the tracer particles are represented in the coordinate system fixed to the advancing meniscus (figs. 3b, 4b). The streamlines are of the same type for both the convec and concave menisci.

From the fact that the liquids flow in the opposite direction on the two sides of the meniscus and that no vortex is found we can conclude that there is no radial flow in the interface. (The statement is, of course, limited for the value of viscosity ratio of the liquids used here.) The above-mentioned fact is of importance in estimating the dynamic interfacial tension and the influence of flow properties on it during the flow in the case of a two-phase system containing surface active agents.

The velocity distributions determined also from the movement of tracer particles contain more information than the flow patterns alone. In addition to streamlines, the axial and radial velocity components versus distance from the meniscus and the tube axis are represented for flat air/water (fig. 5) as well as water/dodecane interfaces (fig. 6). The experimental velocity values are compared with results which had been calculated theoretically by Kafka and Dussan V. [9].

The calculated and measured axial velocity components agree well if $z \geqslant 0.3$ (where z is the axial distance from the meniscus expressed by the tube radius). The radial velocity component has a maximum value near the contact line. The location of that maximum gets closer to the tube axis going away from the interface. The radial velocity gradually becomes negligible and finally vanishes far from the interface. The maximum value and the slope of the measured result surface are smaller than the calculated ones.

Both the calculated and measured values of the axial velocity are smaller and the location of the maximum radial velocity gets closer to the contact line in the case of two-liquid systems than in the air/liquid case due to the higher viscosity of the receding phase (fig. 6) compared to the system of figure 5. This is shown in the flow patterns, too.

It is worth mentioning that the axial velocity distribution at the very vicinity of the interface shows a shape similar to the curvature of the meniscus itself independently of the fact that the interface is flat or curved. The values of the axial velocity components are varying almost independently of the radial position at $z = 0.15$ (see figs. 5 and 6) in accordance with the flat shape of the water/air and water/dodecane

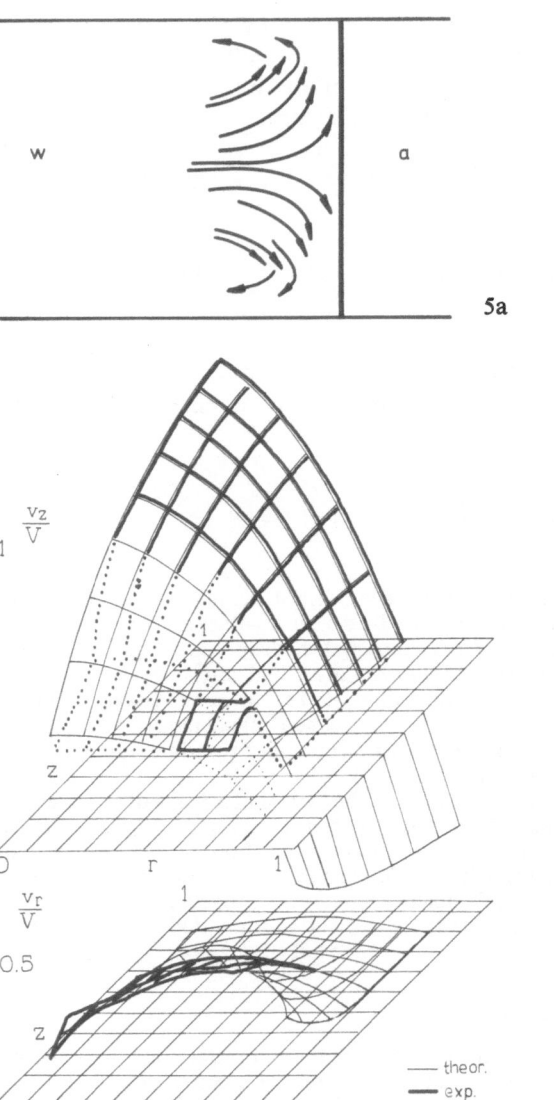

Fig. 5. Flow patterns (a) and velocity distributions (b) of the advancing phase at air displacement by water

interfaces. The profile of the axial velocity distribution at $z = 0.15$ for the case of a concave moving interface (fig. 7) is also adequate to the curved shape of the meniscus except at the wall where the velocity must be decreasing.

Flow patterns and velocity distributions presented here give evidence of the fact, furthermore, that the effect of the meniscus on the tube flow extends over a range of a (the tube radius) on both sides of the interface. In those transition regions in front of and behind the interface the radial flow velocity components are considerable and the axial flow velocity compo-

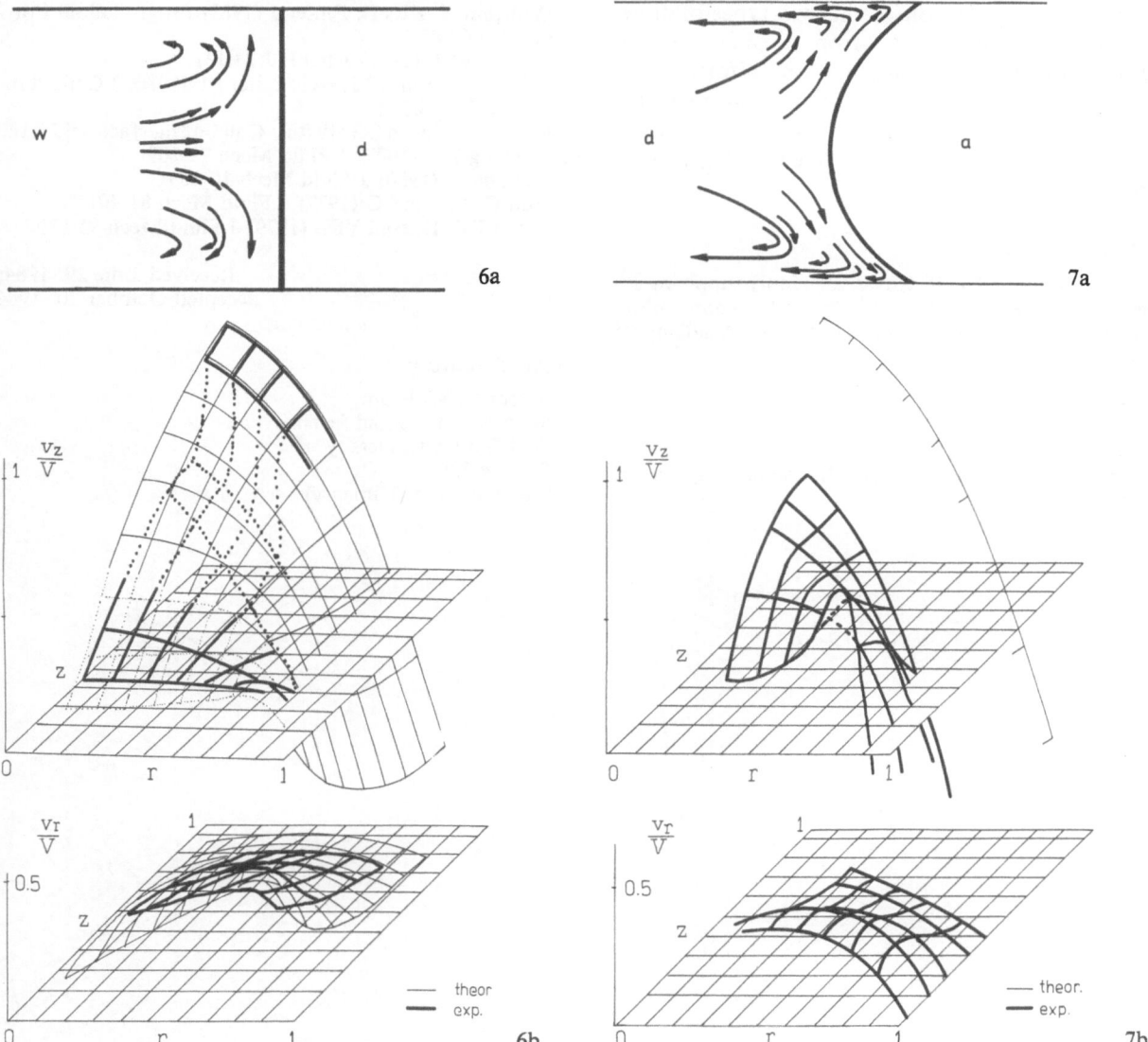

Fig. 6. Flow patterns (a) and velocity distributions (b) of the advancing phase at *n*-dodecane displacement by water

Fig. 7. Flow patterns (a) and velocity distributions (b) of the advancing phase at air displacement by *n*-dodecane

nents show a different distribution and are lower than those of the Hagen-Poiseuille flow. The extension of the transition region does not depend on the shape of the meniscus. This flow behaviour may be important in capillary systems with several fluid interfaces in which the energy loss attributable to the change of the flow at the meniscus can be high.

Conclusions

1. On the basis of flow patterns determined simultaneously in both advancing and receding liquids it has been found that there is no radial flow in the water/dodecane interface during the displacement, whatever its curvature.

2. For the system with flat interface the experimental axial and radial velocity components are in good agreement with the theoretical results of Kafka and Dussan V. except in the immediate vicinity of the meniscus.

3. The experimental axial velocity profile near the meniscus ($z < 0.25$) is adequate to the curvature of the moving interface itself in the cases of flat and curved interfacial shapes as well.

4. The transition region of the two types of flow, viz. Hagen-Poiseuille flow far from the meniscus and the displacement of the meniscus with uniform velocity distribution along the fluid/liquid interface, extends over a range of *a* (the tube radius) on both sides of the moving interface independently of its curvature.

Acknowledgements

We are grateful to Dr. P. Kovás for kindly supplying PS lattices and to the Research Film Centre for cooperation. The work was sponsored by the Hungarian Academy of Sciences.

References

1. Kiss É, Pintér J, Wolfram E (1982) Colloid Polymer Sci 260:808
2. Wolfram E, Kiss É, Pintér J (1983) Progr Colloid Polymer Sci 68:82
3. Dussan VEB (1977) AIChE J 23:131
4. Brown CE, Jones TJ, Neustadter EL (1980) J Colloid Interface Sci 76:582
5. Karnis A, Mason SG (1976) J Colloid Interface Sci 23:120
6. Hocking LM (1977) J Fluid Mech 79:209
7. Lowndes J (1980) J Fluid Mech 101:631
8. Huh C, Mason SG (1977) J Fluid Mech 81:401
9. Kafka FY, Dussan VEB (1979) J Fluid Mech 95:539

Received June 29, 1984;
accepted October 30, 1984

Authors' address:

Professor E. Wolfram
Department of Colloid Science
Loránd Eötvös University
P. O. Box 328
H-1445 Budapest (Hungary)

Progress in Colloid & Polymer Science

Progr Colloid & Polymer Sci 70:119 – 126 (1985)

Some factors affecting the self emulsification of hydrocarbon oils

T. A. Iranloye

Department of Pharmaceutics, Faculty of Pharmacy, University of Ife, Ile-Ife, Oyo State (Nigeria)

Abstract: Hydrocarbon oil-water interfacial tensions have been measured using the detachment of ring method for several systems containing mixtures of surfactants in the aqueous phase. From the plots of $\gamma o/w$ against the concentration of surfactants in solution, slopes i.e. $d\gamma/d\log C$ were determined and arranged in numerical order, the order indicating the magnitude of reduction in $\gamma o/w$ in the presence of the surfactants. The order was subsequently observed to be the same as the order of increasing degree or extent of self emulsification of each hydrocarbon employed.

While studying phase formation when surfactants – hydrocarbon oil mixtures (i.e. three component systems) were added to increasing amounts of water (i.e. forming four component systems) it was observed that those systems which initially developed large quantities of liquid crystalline phases later formed better, finer emulsions than those systems that initially consisted of isotropic phases. These observations and their association with the process of self emulsification or easy emulsion formation in the systems investigated are presented in this report.

Key words: Self emulsification, hydrocarbon oils, factors affecting

Introduction

When two pure liquids are shaken together, droplets of one liquid appear in the other. The droplets soon coalesce and each liquid eventually reverts to its original bulk form which is the thermodynamically stable state. The rapid coalescence of the droplets is due to the high interfacial tension between the two liquids. Addition of a suitable surfactant to one of the liquids results in the formation of a stable dispersion due to the adsorption of the surfactant molecules around the droplets, thus lowering the interfacial tension between the two liquids.

It has long been reported that the energy required for an emulsification process is closely related to the value of interfacial tension, $\gamma o/w$, the higher the value, the greater the energy input required and vice versa [1 – 5]. In certain cases, the tension is so low that emulsification appears to be 'spontaneous' such systems being described as 'self emulsifying' or exhibiting easy emulsification [6 – 10]. Systems possessing low tensions have recently become important in many technologies. These include crude oil recovery, lubrication in the textile industry, lubricating oils for

"cutting" and "rolling" of metals, application of agricultural pesticides and also in the pharmaceutical, cosmetic and paint industries.

A survey of the literature shows that while systems having low tension do self emulsify, other factors can be associated with the phenomenon, of particular importance is the formation of mesomorphic materials [11 – 16]. These states of matter in which the molecular order lies between the almost perfect long-range positional and orientational order found in solid crystals and the statistically long range disorder found in ordinary isotropic amorphous liquids and gases are formed when some surfactants – water and hydrocarbon oils – are mixed in certain proportions under suitable conditions. The simplest approach to obtaining these three components in suitable proportions is by the use of a well constructed ternary phase diagram, the essential features of which have been described at various times in the past [17 – 22].

While conducting an investigation into the effect of formulation variables on the self emulsification of hydrocarbon oils, it was observed that the order of increasing reduction in $\gamma o/w$ in the presence of the surfactants employed as well as the extent of formation

Table 1. Properties of the hydrocarbon oils

Hydrocarbon	Mol. wt.	Viscosity at 25 °C, mPas	Relative viscosity (to water) at 25 °C	γo/w at 25 °C, mNm^{-1}	Chemical composition		
					Aliphatic carbon	Aromatic carbon	Naphtheneic carbon
1. n-Hexane	86.2	0.299	0.33	49.9	+	−	−
2. n-Heptane	100.0	0.397	0.44	50.2	+	−	−
3. Toluene	92.1	0.552	0.62	36.0	+	−	−
4. Cyclonhexane	84.2	0.898	1.00	50.0	+	−	−
5. Dobane*	231.0	10.4	11.64	46.7	+	+	+
6. C$_6$ oil*	225.0	11.2	12.54	43.9	+	+	+

* Pilpel [24]; + present; − absent

of mesomorphic phases could be related to the extent or degree of self emulsification of the selected hydrocarbon oils. The experimental procedures and results are presented below.

Experimental materials

Water was triple distilled from an all glass apparatus (surface tension, γa/w was 71.9 m N m^{-1} at 25 °C).

Hexane, toluene, cyclohexane and heptane were of high purity and were further purified by repeated percolation through beds of Fuller's Earth (adsorption grade, Hopkin and Williams) until their physical properties agreed closely with literature values. Fuller's Earth is known to remove surface active impurities from oils [23].

C$_6$ and Dobane oils, of commercial origin, were similarly purified until constant values of interfacial tension against water, γo/w, were obtained. Other properties of these hydrocarbons are listed in table 1.

The surfactants consisted of mixtures of phosphated nonyl phenol ethoxylate (PNE)/phosphated fatty alcohol ethoxylate (PFE) and Arylan PWS/Ethylan D254. Their properties are indicated in table 2.

Methods

Preparation of three-component "emulsifiable oils" and four-component "emulsions"

The hydrocarbons were added to mixtures of PNE/PFE and Arylan/Ethylan (all preparations by weight) to produce compositions corresponding to different points on three-component phase diagrams as illustrated later.

Four-component systems were prepared by dissolving the surfactants in the hydrocarbons and adding the appropriate amount of water (all preparations by weight %).

Surface (γo/o, a/w) and interfacial (γo/w) tension measurements

Surface tensions were measured with a direct reading du Nouy* [25] tensiometer (Cambridge Instrument Co. *C* 672940) with a platinum ring. All glassware was cleaned with chromic acid followed by thorough rinsing with triple distilled water. The temperature of the water bath was kept at 25 °C ± 0.1 °C. Constant checks of the apparatus were made by measuring the surface tension of the water, γa/w = 71.9 m Nm^{-1} (Harkins [26] gives 72 mNm^{-1}).

Interfacial tensions (γo/w) between the hydrocarbon oils and the aqueous phase (or aqueous solutions of the surfac-

Table 2. Properties of the surfactants

Pairs of surfactants	Type	Molecular weight	Bulk density g cm^{-3}	Viscosity mPas	H.L.B.	Solubility in water	Solubility in the hydrocarbons*
1 PNE	Anionic	677	1.2	40 °C, 11280	−	Soluble	Very slightly soluble
PFE	Anionic	356	1.0	20 °C, 2400	−	Moderately	
				40 °C, 800		Dispersible	Soluble
2 Arylan PWS	Anionic	380	1.0	20 °C, 4318	−	Soluble	Dispersible
				40 °C, 1336			
Ethylan D254	Nonionic	412	1.0	20 °C, 54		Moderately	
				40 °C, 17.0	9.8	Dispersible	Soluble

* See Table 1

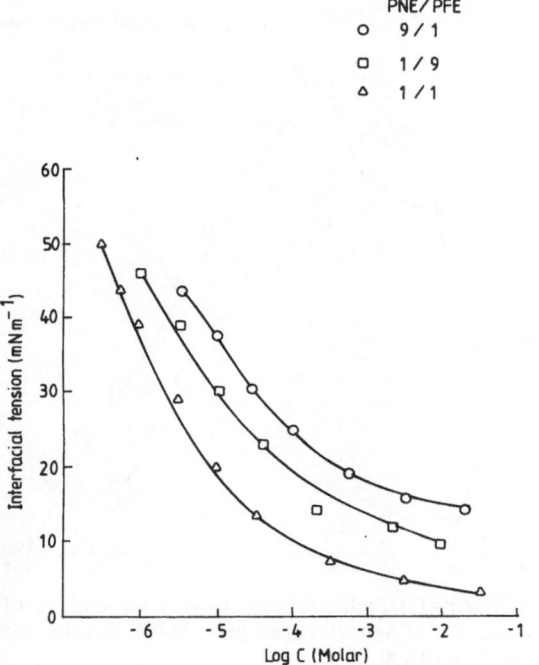

Fig. 1. Interfacial tension/log concentration of surfactants for PNE/PFE/*n*-hexane systems

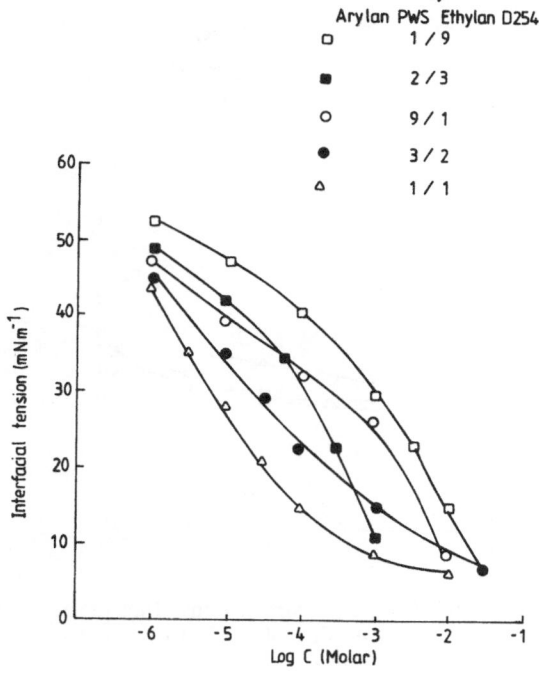

Fig. 2. Interfacial tension/log concentration of surfactants for arylan PWS/ethylan D254/cyclohexane systems

tants) were measured using the same du Nouy tensiometer described above. It was operated according to the A.S.T.M. method [27] and the scale reading at rupture of the interfacial film was converted into interfacial tension by substituting into the following equation:

$$\gamma o/w = P \left[0.725 + \left(\frac{0.0145 P}{C_0^2 (D_1^* - d_1^*)} \right)^{1/2} \right].$$

Here,

$\gamma o/w$ = interacial tension in m N m^{-1}.
P = scale reading at rupture in m N m^{-1}.
C_0 = circumference of the platinum ring, cm.
D_1^* = density of water at 25 °C.
d_1^* = density of the hydrocarbon phase at 25 °C.
(* The values of density were not significantly altered by the very low concentrations of surfactants employed).

Surfactant solutions were prepared as follows. Mixtures of the surfactants in different ratios were diluted with increasing amounts of triple distilled water and the flasks containing the solutions were set aside for some time, drained and filled with fresh solutions of the same strength to compensate for any adsorption of the surfactants by the glass-ware.

About 15 ml of the aqueous phase (or solution) was placed in the du Nouy dish and the hydrocarbon added to a depth of about 0.6 cm. Interfacial tensions were measured after allowing sufficient time (determined from preliminary measurements) for equilibration. After each measurement, the droplets clinging to the ring were shaken back into the

* The choice of this method was purely its convenience and suitability for comparative measurements. Higher accuracies could be obtained by using other methods.

Table 3. Slopes ($d\gamma/d \log C$) mNm^{-1}, derived from interfacial tensions of aqueous solutions of mixtures of PNE/PFE against the hydrocarbons at 25 °C

Surfactants and their ratios		Hydrocarbons					
PNE	PFE	*n*-Hexane	Toluene	Cyclohexane	Heptane	C$_6$ Oil	Dobane
1	1	− 7.8	− 9.0	− 10.0	− 11.3	− 11.8	− 12.3
9	1	− 7.5	− 7.89	− 8.6	− 8.73	− 10.4	− 10.8
1	9	− 6.4	− 8.12	− 8.6	− 8.8	− 9.6	− 11.85

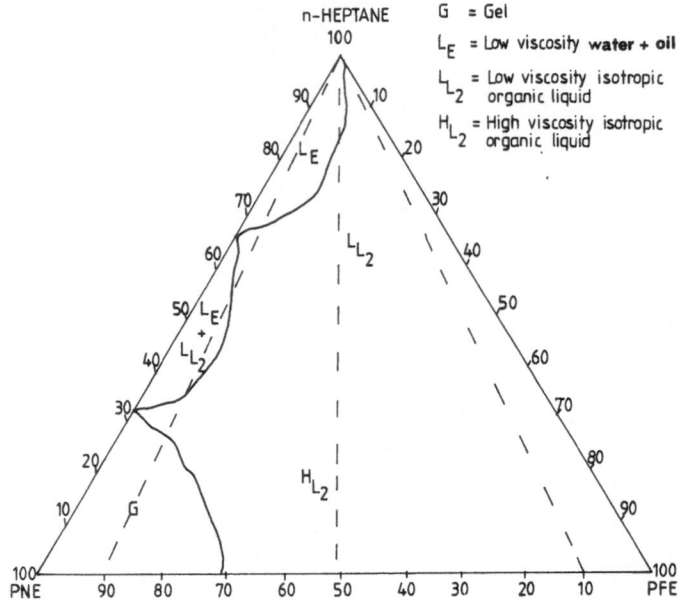

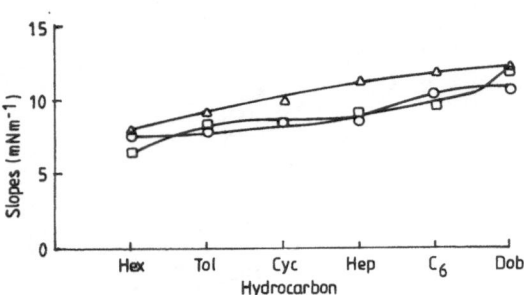

Fig. 3. Slopes ($d\gamma/d\log C$)/hydrocarbons. Interfacial tensions of aqueous solutions of PNE/PFE against selected hydrocarbon oils at 25 °C

Fig. 4. Slope ($d\gamma/d\log C$)/hydrocarbons. Interfacial tensions of aqueous solutions of arylan PWS/ethylan D254 against the selected hydrocarbon oils at 25 °C

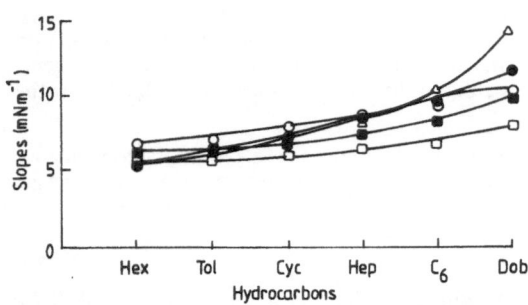

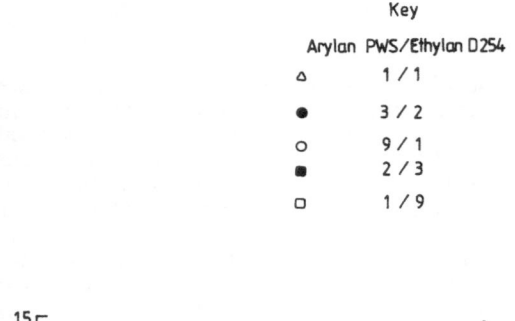

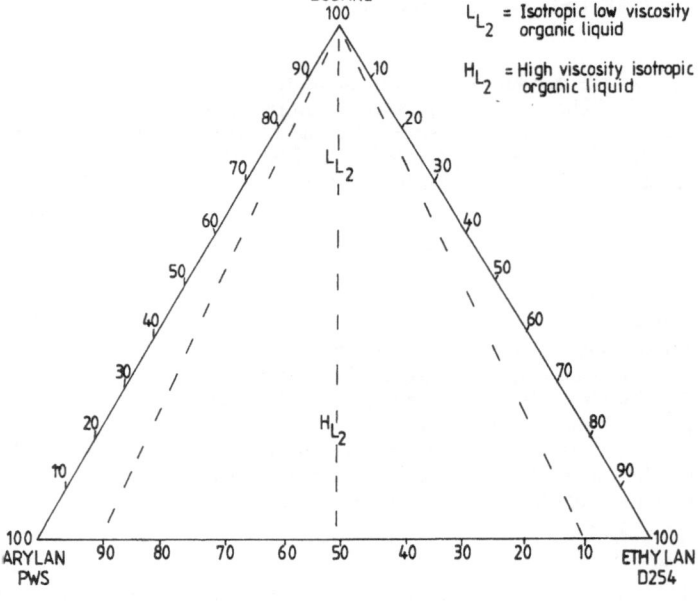

Fig. 5. Phase equilibrium diagram of system containing PNE/PFE/n-heptane at 25 °C

Fig. 6. Phase equilibrium diagram of system containing arylan PWS/ethylan D254/dobane at 25 °C

dish and the ring heated in a Bunsen flame to clean it, keeping the dish covered except during a measurement to minimise evaporation.

Interfacial tensions were obtained as an average of triplicate readings varying by not more than 0.1 m N m^{-1}. It was not possible to measure interfacial tensions <3.0 m N m^{-1}.

Results and discussions

Interfacial tension

The extent of adsorption of the surfactants at the different o/w interfaces was found to be a function of both the pair of surfactants employed as well as the nature of each hydrocarbon oil. Graphs were plotted between interfacial tension (γo/w) and log concentration of surfactants ($\log C$) as illustrated in figures 1 and 2. It is seen from these figures that at low concentrations of surfactants, the graphs of γo/w versus $\log C$ were initially straight lines. The slopes of these lines, i.e. $d\gamma/d\log C$ were therefore determined. similar relationships between γo/w and $\log C$ at low concentrations have been reported for long chain acids, alcohols and ketones [28–31].

It was decided to arrange the slopes in numerical order, depending both on the ratios of the two surfactants in each pair and on the hydrocarbon oils used. A representative result is shown in table 3. The arrangements showed that systems employing PNE/PFE had slopes between −12.3 and −6.4, those containing Arylan/Ethylan had slopes between −14 and −5.5. The different values of the slopes showed that the different surfactants were being adsorbed to different extents at the different interfaces. The higher the absolute value of the slope, the more effective the pair of surfactants was in reducing γo/w.

When the absolute values of the slopes (i.e. ignoring their negative signs) for the various systems were arranged (purely empirically) against the different hydrocarbon oils, they increased regularly from left to right. Each pair of surfactants in each ratio produced a bigger decrease in the interfacial tension of Dobane than of the other hydrocarbon oils. The trend can be seen more clearly in figures 3 and 4 and shows

that the nature of the hydrocarbon as well as of the pair of surfactants affected the values of γo/w.

Phase behaviour

The various phases were observed and identified under polarised light (Allen Viewer, P. W. Allen Co. Ltd.) and by using an optical microscope (Projectina Microscope, Model 4014B, fitted with polarisers, analysers and an Asahi Pentax SP 500 camera) as described by Rosevear [32]. Phase boundaries were located by the triangular staggered technique [33].

Three- and four-component systems

Addition of the hydrocarbon oils to the various surfactant pairs resulted in the formation of different phases. Representative diagrams are illustrated in figures 5 and 6. There was generally a large area of isotropic liquid phase whose viscosity decreased as the amount of hydrocarbon was increased. Some of the three-component systems exhibited a gel region while many also exhibited regions containing emulsions (water + oil) of low viscosity (fig. 5).

The typical phases observed in the four component systems are illustrated in figures 7 and 8. It is seen that those systems containing 10 per cent water formed large areas of isotropic organic liquid phase (L2). This phase consisted of water solubilized within the cores of the inverted micelles of the surfactants in the

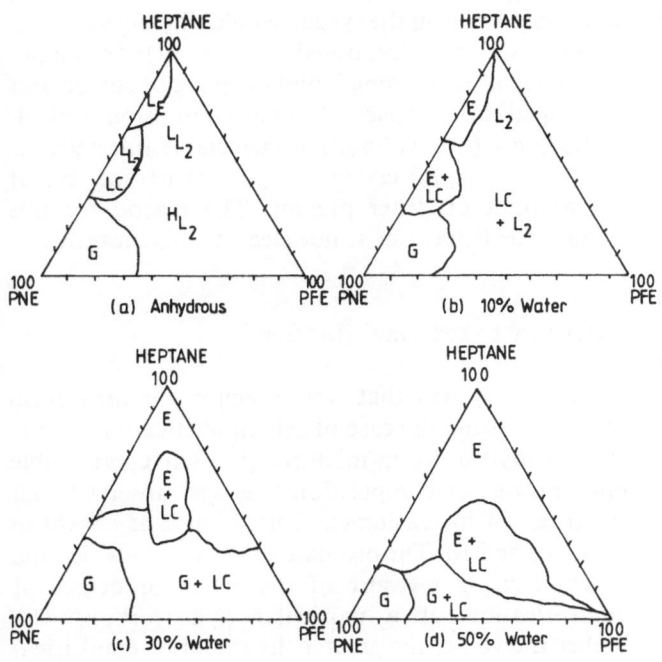

L_{L_2} = Isotropic, low viscosity organic liquid

L_E = Low viscosity **water + oil**

LC = Liquid crystals

H_{L_2} = High viscosity isotropic organic liquid

L_2 = Isotropic organic liquid

G = Gel

E = **oil + water + liquid crystalline phases**

Fig. 7. Phase equilibrium diagrams of system containing PNE/PFE, *n*-Heptane and different amounts of water at 25°C

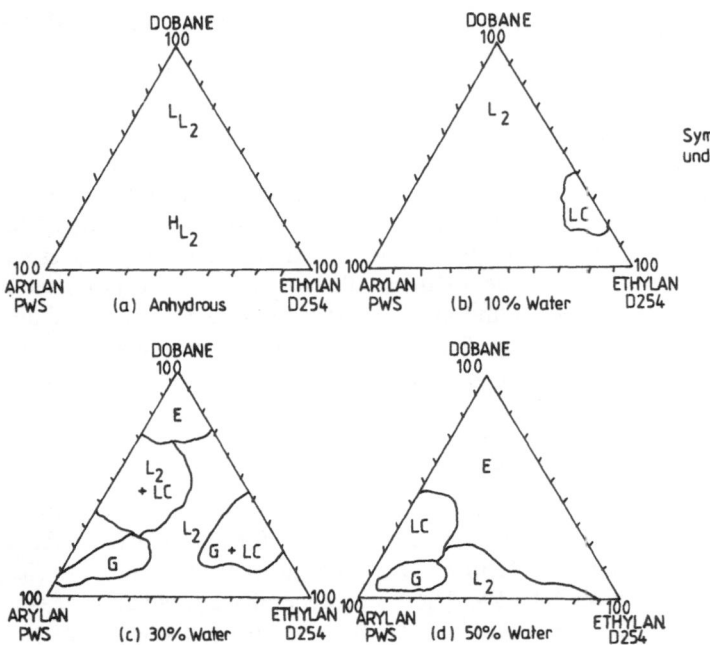

Symbols are as indicated
under Fig. 7

Fig. 8. Phase equilibrium diagrams of system containing arylan PWS/ethylan D254, dobane and different concentrations of water at 25°C

hydrocarbon oils. (In other words, it could be regarded as a micro w/o emulsion). As the concentration of water increased, the L2 phase was generally transformed into either an o/w emulsion (i.e. inversion from w/o to o/w) or liquid crystals or mixtures of liquid crystals and emulsions (i.e. oil + water + liquid crystalline, phases). The phase boundary between the L2 and emulsion phases probably occurred at that surfactant concentration where the inverted micelles were saturated with water.

In the presence of a high concentration of water, e.g. 30%, large areas containing substantial amounts of birefringent liquid crystalline phases were observed in several systems. These observations show that although some liquid crystals developed in the presence of lower concentrations of water (i.e. <30%) large areas consisting essentially of liquid crystalline phases did not form until a significant amount of water was present. This means that an appreciable amount of water is needed for the formation of liquid crystals in these systems, probably as a result of hydrogen bonding. The reason for the differences in extent of formation of liquid crystals among the various systems employing the pairs of surfactants is not clear but it might involve the structural dimensions of each surfactant chain.

Apart from the effect of surfactants on the formation of birefringent materials it was observed that the nature of the hydrocarbon oil and the concentration

of water played a significant role. For example, at low concentration of water (20%), four-component systems employing Arylan/Ethylan, n-hexane and heptane developed large amounts of liquid crystalline materials while viscous gel-like materials developed in a system containing Dobane oil. In the presence of 40% water, they were not detected in quarternary systems containing the former hydrocarbon oils while they were observable in the system employing the latter oil. Toluene was an exceptional oil in that four-component systems containing it and the pairs of surfactants developed large areas of coarse emulsion (oil + water) only (i.e. without any detectable intermediate birefringent liquid crystalline phases) irrespective of the amount of water present. The reason for this anomalous behaviour is not clear at this moment.

Relevance to self emulsification

Using a system that was designed for measuring and compairing the ease of self emulsification of surfactants-hydrocarbon mixtures (i.e. self emulsifiable concentrates or compositions), it was observed that the order of increasing emulsification is as shown in figures 9 and 10. The ordinate axis (i.e. emulsification index) being a measure of the extent or degree of emulsification of a particular hydrocarbon. The higher the value, the greater the extent of emulsifica-

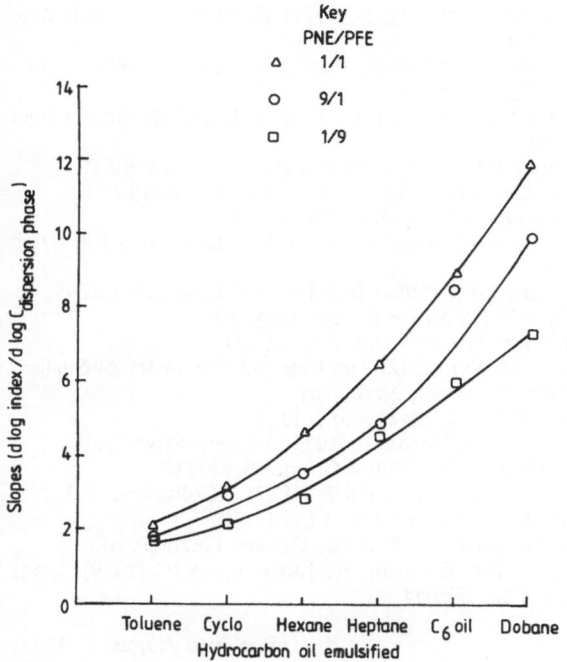

Fig. 9. Slopes (d log index/d log C_{oil}) versus hydrocarbon oil emulsified

tion. From these figures it is seen that the nature of the hydrocarbon oil as well as the pair of surfactants employed influenced self emulsification. Using the pairs of PNE/PFE, Arylan plus/Ethylan D254, the order of increasing emulsification of the hydrocarbon oils was toluene < cyclohexane < hexane < heptane < C$_6$ oil < Dobane. This order is also the order of in-

creasing (or extent of) formation of different networks of birefringent liquid crystalline phases when between 10–50% water was added to the surfactants-hydrocarbon mixtures.

When figures 3, 4 and 9, 10 are compared, it is seen that there is a good correlation between the results of interfacial tension measurement and those of ease of self emulsification, indicating that reduction in $\gamma o/w$ is closely associated with self emulsification in the systems investigated. The slight difference (with respect to the arrangement of the hydrocarbon oils along the abscissa scale) might be due to the random movement of unassociated molecules of surfactants into either the aqueous or oily phases even after prolonged dilution of the emulsifiable oils with water.

Conclusion

The results presented show that using simple methods and materials, it is possible to demonstrate the formation of associated structures and their role in easy emulsion formation. Similarly the part played by reduction in interfacial tension has been presented. The stability of these emulsions in relation to the type and extent of formation of the mesomorphic phases is being investigated and will be the subject of a future communication.

In figures 3, 4, 9 and 10, the hydrocarbon oils were represented as follows:
Hex = hexane; Tol = Toluene; Cyc = Cyclohexane; Hep = Heptane; C$_6$ = C$_6$; Dob = Dobane

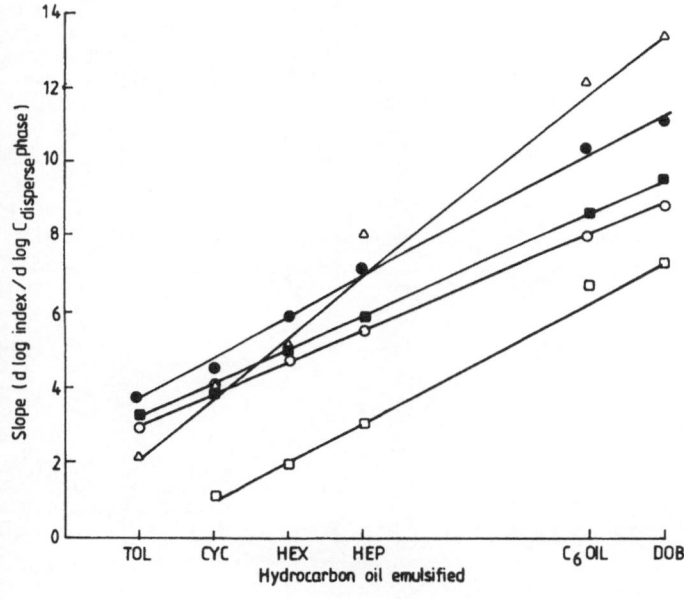

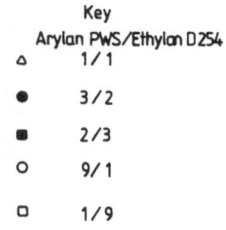

Fig. 10. Slopes (d log index/d log C_{oil}) versus hydrocarbon oil emulsified

Acknowledgements

The author is grateful to Chelsea College, University of London, for use of facilities.

References

1. Quincke G (1888) Wiedemanns Ann 35:593
2. Harkins WD, Zollman H (1926) J Amer Chem Soc 48:69
3. Danielli JF, Harvey EN (1934) J Cell Comp Physiol 5:255
4. Harvey EN, Shapiro H (1934) J Cell Comp Physiol 5:255
5. Christiansen RM, Hixon AN (1957) Ind Eng Chem 49:1017
6. Cayias JL, Schechtec RS, Wade WH (1975) A.C.S. Symposium Series 8, Adsorption at Interfaces, p 234
7. Cayias JL, Schecter RS, Wade WH (1977) J Coll Interface Sci 59:31
8. Wilson PM, Brandner GF (1977) J Coll Interface Sci 60:473
9. Burkowsky M, Marx C (1978) Tenside 15:247
10. Groves MJ (1978) Chem Ind 17, June
11. Friberg S, Mandell L (1969) J Coll Interface Sci 29(1):155
12. Friberg S, Mandell L (1970) J Pharm Sci 59(7):1001
13. Friberg S (1977) Naturwissenschaften 64:612
14. Kislalioglu S, Friberg S (1976) In: South AL (ed) Theory and Practice of Emulsion Technology. Butterworths, London
15. Friberg S, Larsson K (1976) Advances in Liquid Crystals 2. Academic Press, New York
16. Friberg S, Rydhag L (1971) Kolloid-Z u Z Polymere 244:233
17. Ekwall P, Mandell L, Fontell K (1964) J Colloid Interface Sci 29(3):542
18. Ekwall P, Mandell L, Fontell K (1968) Acta Chem Scand 22(1):365
19. Mulley BA (1961) J Pharm Pharmacol 13:205T
20. Rosevear FB (1968) J Soc Cosmet Chem 19:681
21. Gray GW (1966) Mol Cryst 1:333
22. De Jeu WH, Vanderveen J (1977) Mol Cryst Liq Cryst 40:1
23. Murray DRR (1961) Bull Entonol Research 13:205T
24. Pilpel N (1968) Insulation, May, 63
25. du Nouy (1919) J Gen Physiol 521
26. Harkins WD (1952) The Physical Chemistry of Surface Films. Reinhold, New York
27. ASTM Standards (1949) p 1116
28. Ward AFH, Tordai L (1952) J Chem Phys 71:482
29. Hutchinson E (1948) J Colloid Sci 3:219
30. Hutchinson E, Randall R (1952) J Colloid Sci 7:151
31. Pilpel N (1956) J Colloid Sci 11:51
32. Rosevear FB (1968) J Soc Cosmet Chem 19:581
33. Boffey DJ, Collision R, Lawrence ASC (1959) Trans Farad Soc 59:654

Received August 1, 1984;
accepted December 11, 1984

Author's address:

T. A. Iranloye
Department of Pharmaceutics
Faculty of Pharmacy
University of Ife
Ile-Ife
Oyo State (Nigeria)

Subject Index